It was the best of times, it was the worst of times, it was the age of wisdom, it was the age of foolishness, it was the epoch of belief, it was the epoch of incredulity, it was the season of Light, it was the season of Darkness, it was the spring of hope, it was the winter of despair, we had everything before us, we had nothing before us, we were all going direct to Heaven, we were all going direct the other way.

– Charles Dickens, *A Tale of Two Cities*

Development Theory and Practice in a Changing World

Taking a critical and historical view, this text explores the theory and changing practice of international development. It provides an overview of how the field has evolved and the concrete impacts of this on the ground on the lives of people in the Global South.

Development Theory and Practice in a Changing World covers the major theories of development, such as modernisation and dependency, in addition to anti-development theories such as post-modernism and decoloniality. It examines the changing nature of immanent (structural) conditions of development in addition to the main attempts to steer them (imminent development). The book suggests that the era of development as a hegemonic idea and practice may be coming to an end, at the same time as it appears to have achieved its apogee in the Sustainable Development Goals as a result of the rise of ultra-nationalism around the world, the increasing importance of securitisation and the existential threat posed by climate change. Whether development can or should survive as a concept is interrogated in the book.

This book offers a fresh and updated take on the past 60 years of development and is essential reading for advanced undergraduate students in areas of development, geography, international studies, political science, economics and sociology.

Pádraig Carmody is Associate Professor in Geography, Head of Department, Fellow and director of the Masters in Development Practice at Trinity College Dublin and Senior Research Associate at the University of Johannesburg. His research centres on the political economy of globalisation and economic restructuring in Southern and Eastern Africa. He has published in a variety of journals, such as *Economic Geography*, *World Development* and *Political Geography*, amongst others. He has published seven books. The second edition of his *New Scramble for Africa* has recently been published. He sits of the boards of the *Journal of the Tanzanian Geographical Society*, *African Geographical Review*, *Political Geography* and *Geoforum*, where he was previously editor-in-chief. He is currently an associate editor of the journal *Transnational Corporations*, published by the United Nations Conference on Trade and Development.

Development Theory and Practice in a Changing World

Pádraig Carmody

LONDON AND NEW YORK

First published 2019
by Routledge
2 Park Square, Milton Park, Abingdon, Oxon OX14 4RN

and by Routledge
52 Vanderbilt Avenue, New York, NY 10017

Routledge is an imprint of the Taylor & Francis Group, an informa business

British Library Cataloguing-in-Publication Data
A catalogue record for this book is available from the British Library

Library of Congress Cataloging-in-Publication Data
A catalog record for this book has been requested

ISBN: 978-1-138-55177-0 (hbk)
ISBN: 978-1-138-55178-7 (pbk)
ISBN: 978-1-315-14776-5 (ebk)

Typeset in Bembo
by Apex CoVantage, LLC

Contents

Illustrations

Figures

Map

Acknowledgements

This book has been a number of years in the making. I would like to thank Tom Perreault of Syracuse University for suggesting to Routledge that I might be a good person to write it and Andrew Mould for commissioning it. Thanks also to the referees and Andrew Brooks for their detailed comments on the manuscript, which substantially improved it, and President Michael D. Higgins for his endorsement and encouragement of my work. Egle Zigaite's help was invaluable in getting the project through to completion, as was Kate Fornadel's and the copy editor at Apex CoVantage. Parts of chapters draw on previously published work on "States and Development" in *The International Encyclopaedia of Geography: People, the Earth, Environment, and Technology*, D. Richardson, N. Castree, M. Goodchild, A. Kobayashi, W. Liu, and R. Marston (eds.) (London, Wiley Blackwell, 2017); "Matrix Governance and Imperialism" in J. Agnew and M. Coleman (eds.), *Handbook on the Geographies of Power* (London, Edward Elgar, 2018); "Assembling Effective Industrial Policy in Africa: An Agenda for Action", *Review of African Political Economy*, 44(152), 2017, 336–345; "Building BRICS in Africa?", *Handbook of BRICS and Emerging Economies*, P. B. Anand, F. Comim, S. Fennell and J. Weiss (eds.) (Oxford and New York, Oxford University Press, 2019); and book reviews for *The Irish Times* and *the International Development Planning Review*. A section of Chapter 10 also draws on P. Carmody and F. Owusu, "Neoliberalism Urbanization and Change in Africa: The Political Economy of Heterotopias", *Journal of African Development*, 18(1), 2016, 61–73, and I am grateful to Francis (Owusu) for letting me reuse that here. I would like to thank the referees, John Agnew, Mat Coleman, P. B. Anand and Jim Murphy for their comments on one or other of those and the *Proceedings of the National Academy of Sciences of the United States* for permission to reproduce the map of the Millennium Villages. I also thank my son, Daire, for producing Figure 5.1, Ian Yeboah for permission to use a photo (Figure 5.2) from Accra, Sean McCabe for helping with some of the formatting, Howard Stein for statistical advice, the production staff at Routledge and Fiona for her support throughout the process.

Abbreviations

B2B	business-to-business
BPO	Business Process Outsourcing
BRICS	Brazil, Russia, India, China, South Africa
CIA	Central Intelligence Agency (of the United States)
CIVET	Colombia, Indonesia, Vietnam, Egypt, Turkey
CNN	Cable News Network
DAC	Development Assistance Committee (of the OECD)
DfID	Department for International Development (of the United Kingdom)
DRC	Democratic Republic of Congo
ECLAC	Economic Commission for Latin America and the Caribbean
EPRDF	Ethiopian Peoples' Revolutionary Democratic Front
ESAP	Extreme Suffering of African People
EU	European Union
FCCC	Framework Convention on Climate Change (of the United Nations)
FDI	foreign direct investment
G7	Group of Seven
G8	Group of Eight
GA	General Assembly (of the United Nations)
GaWC	global and world cities
GDP	gross domestic product
GPN	global production network
GVCs	global value chains
HIV/AIDS	human immunodeficiency virus/acquired immunodeficiency syndrome
ICT	information and communication technology
ICT4D	information and communication technology for development
IMF	International Monetary Fund
IPCC	Intergovernmental Panel on Climate Change (of the United Nations)
ISI	import substitution industrialisation
MDGs	Millennium Development Goals (of the United Nations)
MINT	Mexico, Indonesia, Nigeria, Turkey

MVP	Millennium Village Project
MVs	Millennium Villages
NAFC	North Atlantic Financial Crisis
NAFTA	North American Free Trade Agreement
NATO	North Atlantic Treaty Organization
NGO	non-governmental organisation
NICs	newly industrialised countries
NUA	New Urban Agenda
ODA	overseas development assistance
OECD	Organisation for Economic Co-operation and Development
OPEC	Organization of Petroleum Exporting Countries
OWG	open working group
PRSPs	Poverty Reduction Strategy Papers
RCTs	randomised controlled trials
REDD	Reduction of Emissions through Deforestation and Forest Degradation
RUF	Revolutionary United Front
SAP	structural adjustment programmes (of the World Bank and IMF)
SDGs	Sustainable Development Goals (of the United Nations)
SEZ	special economic zone
SOEs	state-owned enterprises
STRASA	Stress-Tolerant Rice for Africa and South Asia
TNCs	transnational corporations
TV	television
UK	United Kingdom
UN	United Nations
UNCTAD	United Nations Conference on Trade and Development
US	United States of America
WHO	World Health Organization (of the United Nations)
WTO	World Trade Organization

Introduction

A recent report by the charity Oxfam revealed that the world's richest 42 people shared as much combined wealth as the bottom half of humanity – some 3.7 billion people (Elliot 2018). The 200 biggest transnational corporations (TNCs) globally control the equivalent of nearly a third of the world economy (Weis 2007), while the top 2,000 companies produce about half of the global economic output and are also mostly controlled by rich "white" men. If there is an average of ten people on the boards of each of these, then 20,000 people in the world exercise decisive control over global wealth creation (Harman 2008 cited in Selwyn 2014).

Of the 42 richest people in the world in 2015, 33 were of European extraction ("white") and only 3 were women, all of whom are also white (calculated from Forbes 2015). These (mostly) men enjoy lives of unparalleled power, prestige and privilege. For example, it is reported that in Bill Gates' mansion (the richest person in the world in 2017) the lights and music are programmed to adjust depending on who is in the room. In some cases daily movements in wealth can be eyewatering. For example, the Forbes website now shows the largest variations in wealth for some of the world's wealthiest people on a day-to-day basis. On 10 October 2017 Hui Ka Yan had added US$717 million dollars to his wealth from the previous day, whereas Mukesh Ambani had lost US$729 million, presumably as a result of movement in stock prices. At the same time there are hundreds of millions of people around the world who live in extreme poverty – lacking sufficient food, adequate shelter, clean water and other basic human rights.

While the "billionaire class", as 2016 United States (US) presidential candidate Bernie Sanders called them, have seen massive increases in their wealth in recent decades, from 2010 to 2015 the bottom half of humanity saw a decline in their wealth of US$1 trillion (Hardoon *et al.* 2016 cited in Selwyn 2017). The average income in the world's 20 richest countries is 37 times higher than that in the 20 poorest – a gap that doubled in the space of four recent decades (Wainwright 2008).

The scale of global inequality is vast, repugnant and politically destabilising. For example, the "positive" relationship or correlations between poverty and conflict has been extensively written about (see Collier 2007). Some recent estimates suggest, however, that inequality between individuals around the world is decreasing, on average, with the rise of "emerging economies" such as China. However, these averages hide often widening inequalities within countries and between emerging economies and the least developed ones, for example. Also whether aggregate inequality is reducing is disputed by some who argue that the "Mathew effect" – "to he who has shall be given" – prevails in the global political economy (Wade 2004), or to put it another way, "money makes money" and consequently "the rich get richer and the poor get poorer".

Members of the global political and economic elite, such as former president of the United States, Barack Obama, or Bill Gates argue, that this is the best time in history to be born, given massive progress globally in reducing child mortality, for example, with the number of children dying each year having fallen by more than half from 1990 to 2015, although still at a rate of 16,000 a day, mostly of easily preventable diseases. There has indeed been progress in terms of many social indicators across world regions, although there are disputes as to how these are measured. For example, while many estimates show the number of malnourished people in the world decreasing in recent decades, for those undertaking "intense" physical labour as defined by the Food and Agriculture Organization of the United Nations "the numbers suffering from hunger increased from around 2.25 billion in the early 1990s to approximately 2.5 billion in 2012" (Selwyn 2017, 5). There has, however, been indisputable progress in improving life expectancy around the world, as Figure 0.1 shows, for the world as a whole and select world regions.

As the Commission on Macroeconomics and Health, chaired by Jeffrey Sachs for the World Health Organization, argued in 2001, relatively modest social investments in public health, for example, can have big social and economic pay-offs. However, the discourses of the global elite and other more academic and optimistic accounts of global development, such as Radelet (2015) and Spence (2011), don't take into account that in terms of lived experience there is no such thing as a "global scale". Even if we have to travel for work, as many economic refugees making the crossing from North Africa to Europe need to, we live out our lives in particular places or sets of places and, as intimated earlier, socio-economic conditions and life experiences differ vastly across and within world regions and cities. Consider the massive differences in average life expectancy globally – a reflection of massive global inequality. This is also reflected at local scales.

By way of example, anyone who has travelled to Johannesburg in South Africa, the country and region's commercial capital, cannot but be struck by the glaring inequality in that city – the highest wealth disparity between the rich and the poor of any in Africa – feeding into other social issues such as crime. Income inequality is measured by something called the Gini coefficient. If the Gini coefficient is 1, one person would own everything in the society or consume all the income. If it is at 0, then everyone in the society has the same level of wealth or income. Sometimes the Gini coefficient is set out of 100 instead of 1 but the principle is the same.

Social democratic states with good welfare provision tend to have relatively low Gini coefficients. For example, in terms of income Denmark has a Gini coefficient of 0.25 (World Bank 2017). However, it is far from an equal society, as the richest 10 per cent

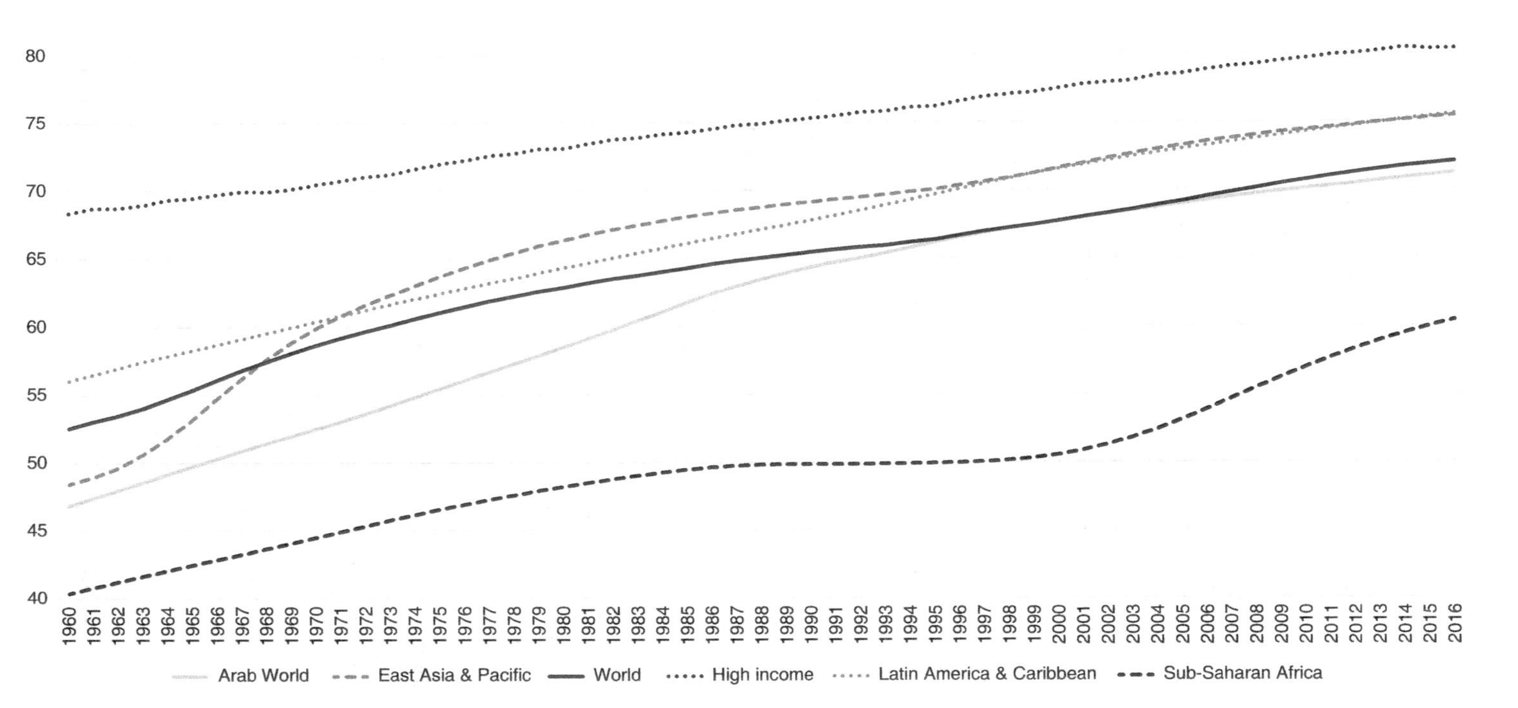

FIGURE 0.1 Life expectancy at birth

of the population have more than five times more income than the poorest 10 per cent (The Local 2015). In Johannesburg the Gini income coefficient is 0.7, and income and wealth are still highly racialised and largely in the hands of people of European descent, given the history of institutionalised racial segregation in that country known as apartheid. In the US the three richest people (Bill Gates, Warren Buffett and Jeff Bezos) own more wealth than the bottom half of the population and in terms of asset ownership, not income, the Gini in the US is over 0.8, a level which is historically associated with political instability, or in the case of the US globalisation backlash, with the election of Donald Trump as president in 2016.

On one visit to Johannesburg when I was giving a talk to a group of ambassadors, I was told the Aston Martin dealership across the road was one of the busiest in the world. There was also a McLaren "supercar" dealership nearby (www.johannesburg.mclaren.com), where the cars sell for hundreds of thousands of dollars. This micro-geography is also replicated in Cape Town, where McLaren and Aston Martin dealerships are side by side in the exclusive Victoria and Alfred waterfront area. Quite why anyone would want to own a luxury car which might make them a target of unwanted attention is curious. However, they are positional goods demonstrating success. While in South Africa I heard a story of a rich "black"[1] South African who said it was his duty to display wealth, through luxury car ownership, to show others it could be done. President Mbeki had previously criticised the country in 2006 for pursuing "personal enrichment at all costs, and the most theatrical and striking public display of that wealth" (cited in Posel, 158 in Death 2016, 127).

In the late 1990s some luxury marque owners in South Africa fitted flamethrowers under their cars to "deter" smash and grabs or carjackers (Figure 0.2). While these devices are relatively rare and not reflective of everyday life in Johannesburg, which functions much the same as in any other major city (Figure 0.3), they do serve as a metaphor, which captures something about the nature of social relations under the "free market" or neoliberalism. There are also of course important international social relations of cooperation, but competition is a hallmark of the current international system. As Yash Tandon (2015) notes, "trade is war".

The exclusion of much of the world's population from the benefits of the current globalised capitalist economic system (those outside the car), despite the fact that much of the wealth of those inside the car is generated by them, represents a form of silent, every day or structural violence (Galtung 1969), which is implicated in the creation of violent conflict, although those "inside the car" also inflict substantial direct violence in other parts of the world – think of the invasion of Iraq in 2003 by the US and Britain.

We might think of those "inside the car"[2] as the "golden billion" (North 2016) living in the rich countries of the Organisation for Economic Co-operation and Development, headquartered in Paris, and those in the global middle class in other parts of the world, although how this might be defined is much disputed, even if they are meant to be a major driver of economic growth in developing countries. The photo is deliberately non-representative and could be argued to be sensationalist, as the vast majority of social interactions in South Africa and elsewhere around the world are not, or at least not directly, violent. However, it is only used as a metaphor to enable us to think through relations of power, different types of violence and how they operate at different scales.

FIGURE 0.2 The "Blaster" flamethrower system

FIGURE 0.3 Street scene in Hillbrow, Johannesburg

While South Africa is often cited as an example where First and Third Worlds come into contact, the global system often spatially separates wealth from poverty, thereby reducing direct conflict but contributing to structural violence and exclusion. While many of the billionaires on the lists of the world's richest people, such as those produced by Forbes (www.forbes.com/billionaires/list/#version:static), made their money through owning high-tech companies, often in the United States, and are physically and socially removed from other parts of the world where there is widespread poverty, is there a relationship or relationships between their very different living conditions and, if so, what are these? One way to think through these issues is with the idea of globalisation.

Globalisation and development

Globalisation has probably been the most written about phenomenon in the social sciences over the last several decades. There are many definitions as to what constitutes globalisation; however, at a basic level, the phenomenon refers to the increased interconnectedness between places, through flows of goods, people or information, for example. While some people argue that globalisation is smoothing out differences between places (cf. Ohmae 1990; Friedman 2007), this is very far from being the case, as the global political economy is characterised by combined and uneven development – that is what happens in one place affects others, producing drastically different and shocking outcomes (Wilber 1973). In fact, globalisation could not take place if everywhere was the same (whatever that might mean in practice), as it is differences between places, such as labour costs, which drive the flows of globalisation, such as foreign investment (Yeung 2002).

Very few things take place at a "global" level in the sense that they affect all places on earth. There are some examples where this is the case, such as climate change or the spread of market-based economies around almost the entire world, with a few exceptions such as North Korea or, to some extent, Cuba. However, other processes happen in particular places or sets of places that are linked up through flows of various types. A t-shirt may link up cotton growers in West Africa with "sweatshop" workers in China and high-income consumers in North America (see Rivoli 2009; Brooks 2015). If the t-shirt is then donated to charity, it may move to a country in Africa, and when sold in a second-hand market it takes on a new "life". A simple commodity such as a t-shirt can then link up many different places, but not everywhere, so some would argue that this represents a form of "translocalisation" (Appadurai 1996) – linking up specific places around the world, rather than globalisation. Although it could also be argued it is a form of globalisation in that it links up different world regions.

Given the highly interconnected nature of the global economy, the types of processes described earlier are also what social scientists call recursive. That is one flow can set off another. The money that is used to pay for a t-shirt from China can be spent buying things from other parts of the world or invested in another country. That, in turn, sets off other countless series of flows, feeding off and generating each other. However, these flows may also generate resistances and backlash and undercut themselves, as was demonstrated during the North Atlantic Financial Crisis (NAFC) of the late 2000s, generating periods of deglobalisation, after which global trade contracted. Some argue that in order to reverse the environmentally and socially destructive effects of transnational capitalism, we need to move towards more purposive, planned and wide-scale deglobalisation (Bello 2013).

Writing in the 1970s Waldo Tobler developed his first law of geography, which stated that "everything is related to everything else, but near things are more related than distant things" (Tobler 1970, 88). However, globalisation has complicated this. For example, there may be more interactions between New York City and other "global" cities, such as London, than with economically depressed parts of upstate New York, for example (Sassen 1991). The "fabulous" wealth of internet billionaires, such as Jack Ma, who founded the Chinese online retailer Alibaba, is partly built on, and facilitated by, electronic devices which contain the mineral coltan, much of which has been extracted in the Democratic Republic of Congo (DRC) under highly exploitative labour conditions (Nest *et al.* 2006). Thus, the "commodity chains" through which much global wealth is produced often impose binding constraints on those who produce value in them, but allow for wealth and freedom for those who plan and control them (Hartwick 1998). The idea behind a commodity chain is that there are different nodes or links in the chain. So, for example, cocoa grown in a locality in Ghana might be one node. Then it is transported to a factory where it is processed, and perhaps made into chocolate, before finally ending up in a shop. So there are different nodes in the chain linked together through transport and logistics.

Despite the inequalities which this globalised system of production produces, it continues to function and further embed and deepen through what the well-known geographer David Harvey (1999) calls "spatial fixes", which disproportionately displace socio-economic and environmental problems to world regions, often in the Global South, with the least power and which had the least role in producing them. For example, Africa is likely to be the continent most adversely affected by global climate change as such a high proportion of the continent's population is dependent on rain-fed agriculture, yet is has produced, in relative terms, very little of the greenhouse gas emissions which are driving this (Toulmin 2009).

Karl Marx famously critiqued what he called commodity fetishism – that is the idea that money establishes equivalences between different types of products or seeming relations between them, obscuring the way in which they are socially produced. The wealth of Europe or the United States is partly built on the importation of cheap products, such as coffee, produced in other world regions with low wages. The reasons these developed parts of the world are wealthy also relate to histories of slavery, colonialism and dispossession. If those of us who live in Europe or North America had to pay more for our coffee or mobile phones, we would be less well off. However, consumers in rich countries don't see these globalised exploitative labour and exchange relations, unless they seek to educate themselves on them, but often only think about the quality and price of the coffee they are buying or how they could spend their money differently (what economists describe as the opportunity cost of consumption – buying something may preclude you from buying something else). Marx marvelled at the way in which capitalism allowed commodities from around the world to be transported and consumed in distant places, but there are also risks associated with this, even for high-income consumers: think of the intensifying impacts of climate change.

With increased globalisation even "societies" far away from where we live are interconnected with our lives through a variety of flows – in that sense we increasingly live in a globalised market society. Increased interdependence was evident in 2017 when UK supermarkets had to ration vegetables in response to the impacts of storms and floods on harvesting in the Mediterranean (Dean 2017), perhaps partly the result of global climate change. "With the UK never more than a few days away from a significant food shortage, UK consumers should also be encouraged over time to reduce how often they eat meat" (Malcolm Bruce, Chair of UKs International Development Committee quoted in Fullfact.org 2013), as its production is more resource and climate intensive than growing crops. Thus, vulnerability not only affects peasant farmers in the developing world, for example, although often with greater impact. Imports of beef or soya from Brazil are implicated in deforestation and dispossession in the Amazon, which also drives climate change in another example of the recursive flows noted earlier.

Whether people are rich or poor, subject to calamitous effects of climate change or relatively insulated from them, depends on social relations, often of globalisation (Buxton and Hayes 2015). As noted earlier, while globalisation is a recursive process, it is also a contradictory one: that is, it tends to create counter-tendencies and what social scientists call counter-movements. The great economic historian Karl Polanyi (1944) talked of double movements towards marketisation and then a backlash against them – such as the global climate justice movement, although this is very heterogeneous in terms of ideological composition (Bond 2012), or the 2016 vote by the United Kingdom to exit the European Union. Sometimes these social movements which seek to resist corporate globalisation, or what is sometimes called globalisation from above, are peaceful and progressive. Think of the Occupy Movement around the world which arose on the heels of the NAFC, sometimes mistakenly called the "global financial crisis" of 2007–2008 even though some economies such as China continued to grow strongly (Routledge 2017). This movement sought greater fairness in the distribution of the world's wealth, away from the so-called 1 per cent, who control most of it, as it has become increasingly concentrated as profits tend to grow faster than economic growth (Piketty 2014). In other cases, these social movements can be violent, nationalistic or exclusionary. Think of the Islamic State in Syria and Iraq or the election of Donald Trump in the United States, despite his anti-Muslim rhetoric and action in the form of a travel ban on several predominantly Muslim countries, for example. The rise of Trump in the United States can, in part, be related to the impacts of the financial crisis on the lives and livelihoods of American voters, in addition to other processes of globalisation associated with the offshoring of manufacturing jobs by TNCs to China and mechanisation and robotisation. Despite being the head of a major corporation, Trump's election rhetoric was anti-corporate in the sense that he critiqued offshoring of production from the US as a practice. However, as president he sought to achieve a new class compromise and reinvigorate the American economy through dramatic reductions in the corporate tax rate. Trump is therefore not anti-corporate, but only against some of their practices (at least rhetorically), even though his daughter Ivanka's clothing line is largely manufactured in Asia.

The world is undergoing major shifts and transformations as a result of the NAFC, the growing role of technology in driving "economic development" and the impacts of climate change. For example, it is estimated that nearly half of all jobs in the US could be subject to automation in the next 20 years (Benedikt Frey and Osborne 2013), with major implications for the rest of the world in terms of patterns of demand, for clothes, for example, mostly produced now in Asia, with China alone producing over half of the world's total (Leonard 2008), although the European Union (EU) is still a significant exporter. The global Sustainable Development Goals (SDGs), adopted in 2015 by the United Nations, are meant to provide a roadmap for humanity to navigate the global challenges of the present and future, but will they survive the rise of populism, which may be either left or right wing and draws its legitimacy through reference to "the people", in North America and Europe? What will the impacts of this conjuncture be on the peoples of the Global South,[3] and how can they have more influence in shaping their destinies?

Structure of the book

Taking a critical and historical view, this book explores the theory and changing practice of international development. It provides an overview of how the field has evolved and the concrete impacts of this on the lives of people in the Global South. It engages the major theories of development, such as modernisation and dependency, in addition to anti-development theories such as post-modernism and decoloniality. It also examines the changing nature of immanent (structural conditions) development, in addition to the main attempts to steer them; what some call imminent development (Cowen and Shenton 1996).

Some argue the era of development as a hegemonic or dominant idea and practice may be coming to an end, at the same time as it appears to have achieved its apogee or height in the SDGs. It may be coming to an end as a result of the rise of ultra-nationalism around the world from India to the US; the increasing importance of securitisation, where powerful actors try to militarise and insulate their lifestyles against those who are excluded; and the existential threat posed by climate change. Whether development will, can or should survive as a concept is also addressed in the conclusion.

Chapter 1 traces the origins of the idea of development in the late colonial period and its emergence as a – if not the – defining ideology of international relations in the post–World War II period. It adopts a historical, but distinctive, approach by examining the nature of the ideology of development, with a particular focus on modernisation theory, and situating it as a theory of international relations, as opposed to a depoliticised, economistic and supposedly value-neutral concept. Part of its utility was its seemingly innocuous nature. Who could be against development and modernisation? Even though the theory of modernisation has been largely discredited theoretically, its underlying premises continue to inform mainstream development practice. It consequently represents a form of "ghost geopolitics" and economics dating from another era but continuing to influence current policies, views, education and other spheres.

The "failure"[4] of much of the "developing world" to modernise and "catch up" with industrial countries in the post-war period led to a variety of alternative approaches to global development. These included the Prebisch-Singer thesis around the generally declining prices for primary commodities and raw materials, Latin American structuralist economics and dependency and Neo-Marxist approaches. Structuralist economics differs from conventional economics, sometimes called monoeconomics, which holds that the same economic laws operate globally. Rather, structuralist economics asserts that it is important to examine the nature of the global economy and the particular structures of the economy being examined, such as whether or not it is dependent on primary commodity exports or not, in designing appropriate policies. Chapter 2 reviews the nature, content and origins of these theories, in addition to interrogating the practice of alternative development based on them through brief case studies of the formerly "socialist" countries of Tanzania and Vietnam and their subsequent reforms.

The coming to power of Margaret Thatcher in the United Kingdom and Ronald Reagan in the United States fundamentally changed the context of international development. Their neoliberal, or free market, ideology coincided with a period of global economic crisis associated with falling rates of profit in the core countries, the second oil shock and the Third World debt crisis. Chapters 3 and 4 explore the theory behind, and implementation of, neoliberalism as it relates to international development. Paradoxically in the same way as under the reign of Queen Victoria in Britain women in the colonies in Africa and elsewhere experienced systematic disempowerment, the neoliberal revolution, led by Thatcher, arguably had a similar impact across much of the developing world. However, at the same time a new school of gender and development was also emergent within the field, which examined the ways in which gender is socially constructed, with important developmental impacts. The relationship between gender and neoliberalism is also explored in these chapters.

Neoliberalism, much like modernisation theory, has been largely discredited in much of the world, partly as a result of the NAFC. The economic theory behind it makes a host of unrealistic and unworkable assumptions, such as that all economic actors have full (perfect) information – that is, that they are all knowing, in the way that a God might be – and that there is no unemployment (Sheppard 2016). However, it has successfully, at least in terms of its own perpetuation, reinvented itself in the developing world through the adoption and co-optation of theories of community participation (Cooke and Kothari 2001), or what has sometimes been called "participatory poverty". Legitimating austerity and programmes of sometimes extreme economic restructuring through recourse to the idea of community has been quite successful in disarming resistance to neoliberalism.

The reproduction of neoliberalism has also been aided by the deepening importance of global value chains (GVCs), similar to the idea of commodity chains referenced earlier, in trade, which are now "hard-wired" into the operation of the global economy and consequently represent a substantial part of its infrastructure. However, global climate change or restructuring, the vote by the UK to leave the European Union (Brexit) and Trumpism may represent an inflection point in international development, similar to the Reagan/Thatcher disjuncture, as even more economic protectionist policies are being adopted by the Trump administration in particular. In a sense this is reflective of a global power struggle for dominance with China – a "smokeless" war for the Global North, or who will be the dominant power in the twenty-first century.

"Free trade", where there are meant to be minimal barriers to trade in terms of border taxes/tariffs, has arguably been more honoured in the breach by the countries of the Global North for products coming from the Global South through policies such as tariff escalation, where raw material imports are subject to minimal import taxes, but these then increase quickly the more processed the goods are. Countries in the Global North want to protect their privileged positions in the global political economy. More protectionist policies in (parts of) the Global North which have been adopted in recent years may speed the adoption of the technologies of the so-called fourth industrial revolution, such as robotisation, which may preclude the need or push towards more carbon-intensive GVCs to source cheap labour, where components and commodities are transported around the world. More protectionist policies raise the costs of imports, making it relatively more profitable to produce in the home market, at least temporarily. However, tariffs on imported raw materials and intermediate goods, such as steel, may decrease profits for their end users. For example, recently enacted steel tariffs in the US make auto manufacturers there less profitable, while raising profits for domestic steel producers. It is the balance between winners and losers which determines their developmental impacts.

Dependency theory, which argues that the wealth of the Global North is partly built on the exploitation of the Global South, producing poverty there, is a heterodox or non-mainstream approach largely developed in and in relation to Latin America, and also Africa, in the 1960s and 1970s (see for example Cardoso and Faletto 1979; Rodney 1973). However, both dependency and neoliberal approaches came under sustained criticism from the so-called Neo-Weberian school in the late 1980s and 1990s.

Max Weber was a famous German sociologist who argued for the importance of bureaucratic rationality in the economic rise of Western Europe, along with other, more dubious factors such as the "Protestant work ethic". Neo-Weberian approaches showed how "autonomous" states in East Asia were able to escape underdevelopment through the adoption of heterodox yet business- or capital-friendly policies. More recent political economy work has analysed the development of the newly industrialised countries (NICs) in Asia from a geopolitical perspective, such as "systematic vulnerability theory", which posits that states may be galvanised to diversify and develop their economies by a security threat (Doner *et al.* 2005). Furthermore, the Neo-Weberian school has come under criticism from those who emphasise the importance of post-war American administration in Asia in the genesis of "developmental states" (Glassman and Choi 2014). Chapter 5 interrogates the theory of the "developmental state" and contrasts it with more geopolitically informed theories. Whether developmental states exist currently in the "developing world" is explored through a case study of Ethiopia.

Beginning in the late 1980s some scholars began to question the development enterprise in its entirety, viewing it as a "Trojan horse" for the assertion of Western interests in the Third World, drawing on post-modernist theory. The term "Third World" is a controversial one, which some people feel is politically incorrect or insulting to these countries, as it is seen to imply a hierarchy where First World is best and Third worst.

However, there a number of things to remember. First the term was initially a geopolitical one for those countries that were neither aligned with the capitalist First World or socialist Second World during the Cold War. Second, the term captures something of the massive power inequality between different parts of the world, as I was taught while I was a graduate student in the US.

Post-modernism questions the ontological (what is real) assumptions of modernism and rationalism that claim there is only one correct way to view the world – through the lens of science. More recently, this has been extended through the idea of decoloniality. Chapter 6 explores the origins, arguments and foundations of post-development theories and examines, through case studies of the Zapatistas in Mexico and related ideas around *buen vivir* in Latin America, their application in practice. The term Latin refers to the dominance of the Spanish and Portuguese languages in that part of the world, but some indigenous people dispute the name America, named for the European Amerigo Vespucci, and instead refer to North America as Turtle Island, for example. The book then moves on in the next section to examine practices of development in more detail.

The year 2015 marked the adoption of the SDGs globally, with 2016 being their first year of operation. As noted earlier there are 17 of these "Global Goals", which aim to both eliminate poverty and achieve worldwide sustainability. These goals and their predecessors – the Millennium Development Goals (MDGs) adopted in 2000 by the UN – were meant to mark the beginning of a new era of globally cooperative development. However, 2016 was also marked by the victory of the forces of globalisation backlash, at least in the US, with the election of Donald Trump, and in the UK, with the Brexit vote. In his inaugural presidential address Donald Trump announced that his focus would be "America first". These events will mark the international development landscape into the future through practices of overseas development assistance (ODA), for example. In the populist framing development aid is only justified when it brings benefits to the donor (Jakupec 2018). When Trump met the Nigerian president, Buhari, in Washington in 2018, he was at pains to note there were barriers to US exports there and that Nigeria "owed" it to the US to reduce these, given the aid disbursed to that country.

Chapter 7 explores the theory and practice of ODA. It interrogates the theory of aid and anti-aid approaches. It also examines the nature of recent theories and practices such as "randomised controlled trials" (RCTs), originally developed in medicine, and their application and implications for ODA. South–South theories of development cooperation are also explored, as is the celebratisation of development, where people such as the Irish rock singer Bono play a highly visible role, celebrated by *Fortune* (2016) magazine as one of the world's greatest leaders for his campaigning work.

Humanity is increasingly told that we live in a global "informational economy". Some non-governmental organizations (NGOs) and multilateral development institutions also extensively promote the idea of information and communication technology (ICT) as a major contributor to, and sometimes "silver bullet" for, international development. Chapter 8 critically interrogates the ideas of information and communication technology for development (ICT4D). It explores the ways in which these theories have developed and evolved and the class interests behind them. The ways in which

ICTs may play a role in structural economic diversification or transformation is also explored.

There is a long history to the idea of "African exceptionalism" in the social sciences. In quantitative studies many economists have found a negative effect on economic growth of being a country in Africa. They have even developed a term to express this – the so-called "Africa dummy" (Englebert 2000). In economics a "dummy" is a term for the unexplained variation in data when other factors are "controlled" for or the data is standardised based on things such as level of education of the population. Englebert explains this dummy by reference to the "arbitrary" nature of post-colonial states in Africa, which relates to the history of African state formation, where European powers largely divided the continent for colonisation by drawing lines on maps at the Berlin and subsequent Brussels conferences. This explains why roughly a third of African borders follow lines of latitude and longitude.

Probably the most influential theory in the literature in relation to the exceptionalism of African development has been the theory of resource curse. While this theory is sometimes validated or contested through so-called large N (number) quantitative studies of dozens of countries, it is often regionalised to Africa and often with a particular focus on the Niger Delta in Nigeria, where that country's oil is produced, with devastating social and environmental effects. Chapter 9 explores the theory behind and practice of the "resource curse". It critically interrogates it through the lens of a transnational rather than a methodologically nationalist ontology. What does this mean? An example of methodological nationalism would be to say that Nigeria is poor because it is corrupt. However, this neglects the ways in which Nigeria as an entity and state has been historically constructed through processes such as colonialism and globalised resource extraction: that is, the way in which local and national resource curses are constructed through transnational networks of actors, or what are known as assemblages (Siakwah 2017a). The location of this chapter may seem peculiar as this section of the book deals with development practice; however, we can think of the resource curse as a form of development practice.

Humanity in the twenty-first century is for the first time a primarily urban species. What are the implications of the "urban era" for international development? Chapter 10 examines the urban transition currently underway across the Global South and its implications for living standards, lifestyles and economic diversification. It also examines older theories, such as the dual economy and parasitic versus generative urbanism debate. It further examines new urban theories, such as those which focus on slumification and planetary urbanisation, to explore the evolution of urbanism in the Global South.

Some theorists have claimed that climate change will undo any development gains which have been accomplished in recent decades and potentially put them into reverse. This may be speeded by the administration of Donald Trump, after he announced he was taking the US out of the United Nations' climate change agreement signed in Paris in 2015. His threats to dramatically cut back American ODA will also potentially affect projects to mitigate (offset) or adapt to the effects of climate change. The election of Trump also signals a new era of strengthened international power politics, where force decides, and a potential reduction in the importance of development as both a discourse of international relations and as a practice, although there are also counter-tendencies as will be discussed. There are numerous other theories and ideologies circulating which serve as counterpoints to these developments, such as ideas around green growth, or the "Beijing Consensus" (Ramo 2004; Halper 2010), which envisages a more active role for the state in guiding the development project and process. This idea plays off of, and serves as counter-point to, the idea of the "Washington Consensus" of free-market economic reforms developed by the economist John Williamson. Some, however, dispute the idea that a Beijing Consensus exists in anything like the same way, as the Chinese government is not overtly prescriptive about the economic policies of its partner countries, and its own economic policies have been pragmatic rather than ideological. Nonetheless, some have argued that as the United States adopts a more internationally isolationist stance, this has opened up space for China to become the de facto leading state in the world. China has also successfully transitioned from being a predominantly agrarian society 40 years ago, to a largely urban-industrial one now. Chapter 11 examines the prospects for rural development in the changed global geopolitical and environmental contexts, which are still influenced by the mega- or meta-trends of financialisation and informationalisation. This is then followed by a short concluding chapter.

This book necessarily reflects my own training and background. In a field or discursive space as vast as international development, it is only possible to cover a small fraction of the available theory and materials. In a book such as this it is not possible to cover all of the potentially important topics in the depth they deserve. For example, the HIV/AIDS (human immunodeficiency virus/acquired immunodeficiency syndrome) pandemic in parts of Eastern and Southern Africa has not been dealt with, even though this has led to the death of millions of people. This has had very serious human and development consequences in the most severely affected countries and for particular social groups. Multiple books have been written just about this; however, space constraints prohibit an investigation of its impacts here. In fact, the whole area of health and development is a massive one but will only be touched on in a few different places during the course of the book. Likewise, there are many other issues that might be worthy of separate chapters, such as gender or education. Gender is dealt with in a number of chapters as a cross-cutting theme, but other topics such as education have been largely omitted, again for reasons of space. Rather, the goal here has been to engage selectively with some/many of the key issues in international development and think about prospects for the future. However, I hope the book provides a useful introduction to some of the main concepts and debates.

My own training is as a critical political economist, with a regional specialisation in Africa, and this is also reflected in the content and approach of the book, although I do also draw on examples from other world regions where appropriate. While most people living in poverty in the world live in Asia, there is a more analytical reason for the focus on Africa. The continent is the poorest in the world, and as the Irish president Michael D. Higgins (2018) notes, it is the crucible where many of the world's greatest challenges, such as climate change and diminishing natural resources, will be played out in the coming century. According to the noted Africa scholar Achille Mbembe (2016) global capitalism is being shaped by events taking place in Africa currently.

My disciplinary background is as a geographer, and this also affects the approach taken. Geographers distinguish between what they call "site" characteristics, or the internal features, and social relations of places and situation, which is how places are positioned relative to or related to other places. More recent work has described what is called "relational place making" – that is, what makes London the city it is, is largely its relations with other parts of the world through flows of finance or people, for example. This geographic way of thinking about the world is particularly applicable or useful when thinking about international development. As will be described later, many orthodox or common theories of development are "internalist" – that is, they ascribe development outcomes based on internal site characteristics of places. An example might be the claim that Africa is poor because it is corrupt or has poor governance. However, externalist or more relational approaches would dispute such an understanding by looking back at the history of European slavery and colonialism on the continent and how that has affected economies and patterns of governance and current international relations there. Understanding how "site" characteristics have been created by interactions with other places is an important insight when understanding patterns of uneven development.

It is also worth making a short comment on the title of the book, which seeks to cover both theories and practices of development. These are often seen as separate fields, with theory being in the realm of the academic or perhaps policy think tanks and government agencies and practice relating to the work of those "in country" or in development organisation offices in other parts of the world. However, the division between theory and practice is, at best, somewhat blurred, as there are intense interactions and overlaps between the two. For example, theory informs practice (or what social scientists sometimes call "praxis" – actions informed by theory) and practice also informs theory. The developing-country debt crisis of the 1980s, discussed in more detail later, led to major changes in which theories received traction in the international development space in addition to changes to theories themselves. Thus, there is no strict separation between theory and practice. For the sake of convenience, the first part of the book is more theoretically focussed and the latter part emphasises practices more, but the two bleed into each other for the reasons discussed earlier. This is a deliberate choice which hopefully makes the reading more enjoyable and informative. Given the focus on practices in the latter part of the book, the inclusion of a chapter on the "resource curse" may seem anomalous, but it can be viewed as a practice of development and is included here for that same reason as noted earlier.

Notes

1. Race is a largely colonial construct designating people by their skin colour.
2. This analogy has been made before by the ecologist Homer-Dixon, cited in the highly controversial and sensationalist piece "The Coming Anarchy: How Scarcity, Crime, Overpopulation, Tribalism, and Disease Are Rapidly Destroying The Social Fabric of Our Planet" (Kaplan 1994), but does not imply acceptance of any of the arguments presented in that article.
3. The term Global South is a contentious and debated one. In broad terms it implies having a post-colonial history, having substantial poverty rates and being "geographically South" – located

in or close to the tropics and subtropics (Peter Dannenberg in conversation, July 2017). There are, of course, exceptions, such as Singapore, which is a highly developed country, even if it does not have any minimum wage across the economy.

4 Sometimes, however, policies succeed through failure. For example, the structural adjustment programmes foisted on many countries of the Global South by the World Bank and the International Monetary Fund were successful in opening markets for Western economies around the world, even as they often had devastating economic and social effects in the countries in which they were implemented (see for example Onimode 1989).

References

Appadurai, A. 1996. *Modernity at Large: Cultural Dimensions of Globalization.* Minneapolis, MN and London: University of Minnesota Press.

Bello, W. 2013. *Capitalism's Last Stand? Deglobalization in the Age of Austerity.* London: Zed Books.

Benedikt Frey, C., and M. Osborne. 2013. *The Future of Employment: How Susceptible Are Jobs to Computerisation?* www.oxfordmartin.ox.ac.uk/downloads/academic/The_Future_of_Employment.pdf.

Bond, P. 2012. *Politics of Climate Justice: Paralysis Above, Movement Below.* Scottsville, SA: University of Kwazulu-Natal Press.

Brooks, A. 2015. *Clothing Poverty: The Hidden World of Fast Fashion and Second-Hand Clothes.* London: Zed Books.

Buxton, N., and B. Hayes, eds. 2015. *The Secure and the Dispossessed: How the Military and Corporations Are Shaping a Climate-Changed World.* London: Pluto.

Cardoso, E., and F. Faletto. 1979. *Dependency and Development in Latin America.* Berkeley, LA and London: University of California Press.

Collier, P. 2007. *The Bottom Billion.* Oxford and New York: Oxford University Press.

Cooke, B., and U. Kothari. 2001. *Participation, the New Tyranny?* London: Zed Books.

Cowen, M., and R. Shenton. 1996. *Doctrines of Development.* London: Routledge.

Dean, S. 2017. "What Is Causing the 2017 Vegetable Shortage and What Does it Mean for Consumers?" *The Telegraph.* www.telegraph.co.uk/business/2017/02/03/causing-2017-vegetable-shortage-does-mean-consumers/.

Death, C. 2016. *The Green State in Africa.* New Haven and London: Yale University Press.

Doner, R.F., B.K. Ritchie, and D. Slater. 2005. "Systemic Vulnerability and the Origins of Developmental States: Northeast and Southeast Asia in Comparative Perspective." *International Organization* 59 (2): 327–61.

Elliott, L. 2018. "Inequality Gap Widens as 42 People Hold Same Wealth as 3.7bn Poorest." *The Guardian.* www.theguardian.com/inequality/2018/jan/22/inequality-gap-widens-as-42-people-hold-same-wealth-as-37bn-poorest.

Englebert, P. 2000. "Solving the Mystery of the Africa Dummy." *World Development* 28 (10): 1821–35.

Forbes. 2015. "Forbes Billionaires: Full List of the 500 Richest People in the World." www.forbes.com/sites/chasewithorn/2015/03/02/forbes-billionaires-full-list-of-the-500-richest-people-in-the-world-2015/#38fbc7445b92.

Fortune. 2016. "Why U2's Bono Is One of the World's Greatest Leaders." http://fortune.com/bono-u2-one/.

Fullfact.org. 2013. "Is the UK's Food Supply Hanging in the Balance?" https://fullfact.org/economy/uks-food-supply-hanging-balance/.

Galtung, J. 1969. "Violence, Peace, and Peace Research." *Journal of Peace Research* (3): 167–91.

Glassman, J., and Y. Choi. 2014. "The Chaebol and the US Military-Industrial Complex: Cold War Geo-Political Economy and South Korean Industrialization." *Environment and Planning A* 46 (5): 1160–80.

Halper, S. 2010. *The Beijing Consensus: How China's Authoritarian Model Will Dominate the Twenty-First Century.* New York: Basic Books.

Hartwick, E. 1998. "Geographies of Consumption: A Commodity-Chain Approach." *Environment and Planning D-Society & Space* 16 (4): 423–37.

Harvey, D. 1999. *The Limits to Capital.* New edition. London: Verso.

Higgins, M.D. 2018. "Africa-Ireland Relations: Current and Future." *Speech to the Trinity International Development Initiative*, May. www.president.ie/en/media-library/speeches/africa-ireland-relations-current-and-future.

Jakupec, V. (2018). *Development Aid – Populism and the End of the Neoliberal Agenda.* Cham, Switzerland: Springer.

Leonard, M. 2008. *What Does China Think?* New York: Public Affairs.

The Local. 2015. "Denmark has OECD's Lowest Inequality." www.thelocal.dk/20150521/denmark-has-lowest-inequality-among-oecd-nations.

Mbembe, A. 2016. *Africa in the New Century.* http://africasacountry.com/2016/06/africa-in-the-new-century (last accessed 11 September 2018).

Nest, M., F. Grignon, and E. Kisangani. 2006. *The Democratic Republic of Congo: Economic Dimensions of War and Peace.* Boulder, CO and London: Lynne Rienner Publishers.

North, D. 2016. *A Quarter Century of War: The US Drive for Global Hegemony.* Oak Park: Mehring Books.

Ohmae, K. 1990. *The Borderless World: Power and Strategy in the Interlinked Economy.* London: Fontana.

Onimode, B. (ed.) 1989. *The IMF, The World Bank and the African Debt: The Economic Impact.* London and New Jersey: Zed Books.

Piketty, T. 2014. *Capital in the Twenty-First Century.* Cambridge, MA: Harvard University Press.

Polanyi, K. 1944. *The Great Transformation: The Political and Economic Origins of Our Time.* Boston: Beacon Books.

Radelet, S. 2015. *The Great Surge: The Ascent of the Developing World.* New York: Simon and Schuster.

Ramo, J. 2004. *The Beijing Consensus.* London: Foreign Policy Centre.

Rivoli, P. 2009. *The Travels of a T-Shirt in the Global Economy: An Economist Examines the Markets, Power, and Politics of World Trade.* 2nd edition. Hoboken, NJ: Wiley Chichester; John Wiley [distributor].

Rodney, W. 1973. *How Europe Under-Developed Africa.* Paris: Bogle-L'Ouverture.

Routledge, P. 2017. *Space Invaders: Radical Geographies of Protest. Radical Geography.* London, UK: Pluto Press.

Sassen, S. 1991. *The Global City: New York, London, Tokyo.* Princeton, NJ and Oxford: Princeton University Press.

Selwyn, B. 2014. *The Global Development Crisis.* Cambridge: Polity.

———. 2017. *The Struggle for Development.* Cambridge and Malden, MA: Polity.

Sheppard, E. 2016. *Limits to Globalization: Disruptive Geographies of Capitalist Development.* Oxford and New York: Oxford University Press.

Siakwah, P. 2017. "Are Natural Resource Windfalls a Blessing or a Curse in Democratic Settings? Globalised Assemblages and the Problematic Impacts of Oil on Ghana's Development." *Resources Policy* 52: 122–33.

Spence, M. 2011. *The Next Convergence: The Future of Economic Growth in a Multi-Speed World.* Crawley, WA: UWA Publishing.

Tandon, Y. 2015. *Trade Is War: The West's War Against the World.* New York and London: OR Books.

Tobler, W. 1970. "A Computer Movie Simulating Urban Growth in the Detroit Region." *Economic Geography* 46: 234–40.

Toulmin, C. 2009. *Climate Change in Africa.* London: Zed Books.

Wade, R. 2004. "On the Causes of Widening World Income Inequality, or Why the Matthews Effect Prevails." *International Journal of Health Service* 35 (4): 631–53.

Wainwright, J. 2008. *Decolonizing Development: Colonial Power and the Maya.* Oxford: Wiley-Blackwell.

Weis, A. 2007. *The Global Food Economy: The Battle for the Future of Farming.* London: Zed Books.

Wilber, K. (ed.) 1973. *The Political Economy of Development and Underdevelopment.* Swedish, SL: Random House.

World Bank. 2017. "Gini Index (World Bank Estimate)." https://data.worldbank.org/indicator/SI.POV.GINI?view=map&year=2005.

Yeung, H.W.C. 2002. "The Limits to Globalization Theory: A Geographic Perspective on Global Economic Change." *Economic Geography* 78 (3): 285–305.

CHAPTER

1 The idea of development and modernisation

> The first step toward any alternative vision lies therefore in the recognition that African cities are quite different from one another in patterns, processes, forms and functions.
>
> – (Myers 2011, 7)

How we perceive things depends on our upbringing, value sets and education, amongst other factors. Social scientists and others sometimes mistake correlations with causation. In a conversation with a taxi driver recently I was told that South Africa was "very developed" because of the presence of "Europeans" (presumably people of European extraction there). I was somewhat taken aback by this statement, particularly as the taxi driver in question was originally from Nigeria. Then I got to thinking about it and realised that this was probably quite a common, if racialised, way of thinking about the geography of economic development. It is also what Robert Cox (1987) calls a "synchronic" or present-focussed way of looking at the world, without thinking about the histories of slavery, dispossession, colonialism and ultimately the brutal, enforced system of racial segregation in South Africa known as apartheid (the Afrikaans word meaning apartness), which enriched people of European extraction and impoverished people of African ancestry for the most part. Afrikaans is a language spoken in Southern Africa, primarily derived from Dutch, and is spoken by self-defined Afrikaners, which translates as Africans – people primarily of Dutch, French and German origin who identified with their "new homeland", rather than the "old county" or countries – and people of mixed ancestry (so-called "coloureds").

Cox contrasts a synchronic with what he calls a "diachronic" approach, which takes history into account. However, synchronic and racialised explanations are still very common, as people often make associations between what they observe in reality currently without perhaps investigating more deeply. This tendency has serious implications for geopolitics and the practice of international development.

In 2017 the newly elected president of France, Emmanuel Macron, gave a speech in which he said Africa had "civilizational problems" (quoted in Anyangwe 2017). As intimated earlier, this way of thinking is perhaps somewhat intuitive – "Africa must be poor because of its characteristics" – rather than thinking about the way in which Africa has been relationally produced through its interactions with other places (Pierce *et al.* 2011), in the same way as anywhere else around the world. Indeed, the name for the mega-city, Lagos in Nigeria is from the Portuguese for lake, bespeaking its colonial origins. Likewise, poverty is relationally produced through power and social relations, rather than being natural or residual (Selwyn 2014).

"Internalist" explanations of underdevelopment focus on the characteristics of places and then associate/blame their level of development on them. A common example of this is the argument that "Africa is poor because it is corrupt", which is also what geographers call a meta-geographical claim – generalising a social characteristic to an entire territory. An "externalist" explanation might argue that "Africa is poor because it is exploited by other places" (or people in other places). However, in a sense this internal/external division is somewhat unhelpful, even if it is true that the continent has historically been exploited by outsiders, because places are made relationally, so even things like corruption have been structured and shaped by interactions with other places. For example, corruption is associated with colonialism in Zambia, where the indigenous colonial administrative class (the Boma class) sought to emulate the lifestyle of the colonial bourgeoisie or rulers and generate its own resources, through graft, for business development (Chipungu 1992).

There are many different ways of viewing and explaining the world and its highly uneven geography. Partly how we interpret the world depends on what we choose to pay attention to in our analysis. Many Europeans and Americans view capitalist economic development as a normal process and way of being; perhaps implicitly thinking that if other places around the world "fail" to follow in their footsteps, it must be because they are corrupt, or have "bad geography", with what they consider to be unhealthy climates, for example. This kind of "internalist" thinking, which places the blame for underdevelopment on the internal characteristics of particular places, also conveniently elides or obscures the ways in which Europe and North America have been involved in the creation of poverty in other parts of the world through slavery, colonialism and enforced programmes of market opening. Rather, Europeans and North Americans often see their countries as benevolent "donors" – giving aid to other countries to help them develop.

For a long time it was assumed by many in Europe and North America that other parts of the world would "catch up" with the West and develop economically in their image. However, such a line of thinking, which views history as a kind of conveyor belt, has been shown to be fundamentally wrong. This kind of thinking, which views one thing as inevitably leading to another, is called teleology by social scientists. An example of teleological thinking would be Karl Marx's theory that capitalism would necessarily give rise to communism, which has proven to be false, as capitalism has shown itself to be extremely long-lived, technologically dynamic and adaptable to changed circumstances.

Chakrabarty (2000) has written about an approach to history that he calls "History 1" which is where the European experience of class relations or nationalism is extrapolated as analytical categories to other parts of the world. However, the reality more closely accords to what he calls "History 2", which is the substantially messier actuality of local, national and regional diversity and human interaction, social life, structures and experiences, shaped not only by the seemingly powerful but also by those holding more subaltern or subordinate class, racial, gender, ethnic or national positionalities.

Geographers have recently drawn on this theory. For example, Derickson (2015, 647) discusses what she calls

> "Urbanization 1" and "Urbanization 2" . . . Urbanization 1 is exemplified by the planetary urbanization thesis that posits the complete urbanization of society, whereas Urbanization 2 is characterized by a more diverse set of interventions, united by a political and epistemological strategy of refusing Eurocentrism, thereby "provincializing" urban theory.

Garth Myers in the opening quote for this chapter is getting at a similar idea: that we should be wary of extrapolating homogenising concepts and theories around the world based on the experiences of other places and that we need to be open to, and attentive to, differences.

We can also extend such analysis into thinking about contemporary types of globalisation – Globalisation 1 and 2 – where financial or corporate globalisation "from above" (Korten 1999) (Globalisation 1) may appear as a dominating or homogenising force, whereas in reality there are also a variety of different types of globalisation "from below", such as African traders migrating to China (Lee 2014) or small firms establishing marketing links and exporting internationally (Globalisation 2), even if the two types of globalisation from above and below are interrelated. This diversity offers potential not only for theoretical development but also practice, which can inform theory.

The origins of development

Where we end up then, in terms of our thinking, partly depends on where we start. Many texts on international development begin with a Euro-American focus. President Truman of the United States is often credited with coining the term "development". It was he, in his inaugural speech in 1949, who spoke of the need for the United States to share its great technological prowess with other parts of the world so that they could progress, thereby issuing in the "era of development". According to him:

> More than half the people of the world are living in conditions approaching misery. Their food is inadequate. They are victims of disease. Their economic life is primitive and stagnant. Their poverty is a handicap and a threat both to them and to more prosperous areas. For the first time in history, humanity possesses the knowledge and skill to relieve suffering of these people. The United States is pre-eminent among nations in the development of industrial and scientific techniques. The material resources which we can afford to use for assistance of other peoples are limited. But our imponderable resources in technical knowledge are constantly growing and are inexhaustible. I believe that we should make available

> to peace-loving peoples the benefits of our store of technical knowledge in order to help them realize their aspirations for a better life. And, in cooperation with other nations, we should foster capital investment in areas needing development. Our aim should be to help the free peoples of the world, through their own efforts, to produce more food, more clothing, more materials for housing, and more mechanical power to lighten their burdens. We invite other countries to pool their technological resources in this undertaking. Their contributions will be warmly welcomed. This should be a cooperative enterprise in which all nations work together through the United Nations and its specialized agencies whenever practicable. It must be a worldwide effort for the achievement of peace, plenty, and freedom. With the cooperation of business, private capital, agriculture, and labor in this country, this program can greatly increase the industrial activity in other nations and can raise substantially their standards of living.
>
> (Truman 1949)

However, that is only one place to start, and Truman's speech took place against the backdrop of the emerging Cold War with the Soviet bloc and the search for Western allies in other parts of the world. It was called a Cold War because Soviet and Western troops did not come into direct confrontation, although American and communist Chinese troops did fight each other during the Korean War in the 1950s. However, it took the form of a hot or active shooting war in much of the Global South, where the superpowers often supported competing warring sides in civil conflicts (Shubin 2008). According to the World Bank war is development in reverse, although some argue that (uneven) development is implicated in war as discussed later.

Others argue that the idea of development can be traced back to Aristotle's concept of eudaimonia (human flourishing) or Adam Smith's theory of the different stages of human social development from hunters to commerce (Selwyn 2014). Likewise, starting from China or Niumi in Gambia (Wright 2010) might give us a very different perspective. As China has urbanised and industrialised, rural villages have been absorbed into cities, giving rise to distinctive concepts and practices, such as "village in the city" (Myers 2018). This is not to suggest that Europe or the United States has not had a massive impact on the ways in which other places have developed or been produced; in the same ways as other places have largely shaped how these places have developed (think of the impact of Europe on the history of the US) or the capital which slaves in the Caribbean generated, which was used to propel industrialisation in Britain (Williams 1994). Likewise, "the West" has been partly produced through "innovations" from elsewhere, such as algebra, whose name comes from the title of a book by the Persian mathematician al-Khwarizmi (Stuenkel 2016).

The concept of development could be said to be a superior or colonial one – the idea that some people or places are backward and need to adopt different practices similar to those in Europe or North America for them to catch up with what are often perceived to be better or superior lifestyles. According to Derek Hall (2013, 85) "British imperialism was marked almost from its inception by the idea that taking over land in the colonies for productive purposes was an act of 'improvement' (the original meaning of which was 'to make profitable') that would benefit not just those acquiring the land but all humankind". The arch imperialist, Cecil Rhodes, whose vision of British rule in Africa was that it would extend from Cairo to the Cape in South Africa, which was in fact achieved, conveniently defined the colonial enterprise as "philanthropy plus a 5 percent dividend on investment" (quoted in Khoo 2013, 93), whereas the French referred to their civilising mission, or *mission civilisatrice*, in their colonies. Thus, the reality of exploitation was disguised as concern for others and their betterment. Similar critiques are levelled against the idea of development today. The reality of poverty as expressed through relative powerlessness, hunger, lack of access to adequate shelter and other manifestations is only too real for hundreds of millions of people around the world, largely as a result of colonialism and subsequent waves or rounds of globalisation.

Modernisation theory

Perhaps the most influential theory of development after the Second World War and during the Cold War was modernisation theory. This had both economic and more political strands. The basic idea behind it was that rich Western countries had attained the "good life" for the majority of their populations and that they could show other countries the way to achieve the same for their peoples. This theory was most famously propounded and developed by Walt W. Rostow, who was an economic history professor and also served as special assistant for National Security Affairs under US President Lyndon B. Johnson from 1966 to 1969, in addition to holding a variety of other roles in the US government.

Rostow published his widely read and cited book, *The Stages of Economic Growth: A Non-Communist Manifesto*, in 1960. It is interesting that the highest government office he held was in the realm of national security. Consequently, his book should be seen as a product of the historical context of the so-called Cold War between the Western powers, led by the US, and the Soviet bloc. Indeed, the sub-title to Rostow's book – A Non-Communist Manifesto – hinted at its geopolitical impetus.

Rostow (1960, 1) wrote in the introduction to his book that he was presenting a "way of generalising the sweep of modern history. This form of this generalization is a set of stages-of-growth". The stages he presented were 1) the traditional society, 2) the pre-conditions for take-off, 3) the take-off, 4) the drive to maturity and 5) the age of high mass consumption. This "stagist" theory of development mirrored Marx's theory of history progressing in stages, although the end point for him was different – communism, rather than a high mass-consumption capitalist society envisaged by Rostow. For Rostow, as perhaps for Marx, history operated like a conveyor belt, where societies almost inevitably went through a series of transitions.

In contrast to the ways in which modernisation theory is sometime caricatured, Rostow does not ignore colonialism or power inequality in the international system. However, the impacts are voluntarist, or the idea that human will is the dominant force in history, rather than social structures, according to Rostow. He asks, "How should the traditional society react to the intrusion of a more advanced power: with cohesion, promptness and vigour, like the Japanese; by making a virtue out of fecklessness like the oppressed Irish of the eighteenth century; by slowly and reluctantly altering the traditional society, like the Chinese" (16). There is an element here of blaming the victims and replicating colonial tropes about the lazy or drunken Irish, Africans, etc. Japan was never colonised, even if it was for a period of time subject to unfavourable trade treaties. Likewise China was never fully colonised, even if the Western powers sacked different parts of the country and took various "treaty ports", such as Hong Kong, as their own, after the Opium Wars or "Global War for Drugs", at least for a period of time. In contrast British colonial rule in Ireland was extremely intensive, at times bordering on genocidal. One British economist and sometime government advisor is reported to have said at the time of the great famine that "the famine would not kill more than one million people, and that would scarcely be enough to do any good" (Nassau Senior quoted in Nally 2011, 211). Some saw that famine as an opportunity to restructure Irish agriculture, obscuring the way in which land dispossession of Catholics was largely responsible for the famine in the first instance. Given the fact that China and Japan were not colonised, it is then perhaps not accidental that they are now the world's second and third largest economies, respectively, although it is important to remember that Japan and China have their own imperial histories, so their development was not "self-contained".

The British fought the Opium Wars in China to open up the country to the drug, much of which was produced in the British colony of India, in order to offset their trade deficit with China, largely as a result of tea imports. Subsequently, Britain disavowed the drug trade, but not the other deadly one. According to Robin Hanan, formerly of the European Anti-Poverty Network, there are two trades in the world that kill people – the drug trade and the arms trade (in conversation). The drug trade largely flows South–North and is illegal, whereas the arms trade largely flows North–South but is by contrast largely legal, bespeaking prevailing patterns of power in the global political economy.

A common way of dividing colonies is between settler and non-settler. Settler colonies experience substantial in-migrations of people from Europe, whereas non-settler ones did not and were largely policed and ruled through force and/or co-optation and/or creation of locally cooperative elites. According to Rostow "external forces" could prepare a country for transition to modernisation by establishing the precondition for take-off. In relation to the primary British settler colonies of the US, Canada, Australia and New Zealand, he notes that

> the creation of the preconditions for takeoff was largely a matter of building social overhead capital – ports and roads – and of finding an economic setting in which a shift from agriculture and trade to manufacture was profitable; for, in the first instance, comparative advantage lay in agriculture and the production of food stuffs and raw materials for export.
>
> (1960, 18)

The expropriation and murder of indigenous people is not mentioned. Through history, however, some have celebrated this. For example, Adolf Hitler's armoured train was called "Amerika" up until Germany declared war on the United States, reportedly to celebrate the way in which the indigenous people there had been murdered/exterminated. The role of slavery or colonialism in establishing the "preconditions" for industrialisation in the UK or the US goes unremarked by Rostow – a somewhat surprising omission for a historian, but again perhaps bespeaking a geopolitical impetus in the work, as the United States sought to present an alternative, "positive" vision of society to that of socialism, which was in the ascendancy across much of Asia during that time.

While Rostow's approach was ahistoric it did have some notable seeming successes in practice – in particular, in South Korea and Taiwan where massive US aid was an important part of their industrialisation drive (Amsden 1989; Pempel 1999). Furthermore, the US supported land reform in these countries and the crushing or disbandment of the local landlord class – as the maldistribution of land was seen to be a source of peasant discontent and revolt in Asia, for example, in Vietnam. However, there was nothing "natural" about the way in which "take-off" occurred in these societies – contrary to Rostow's vision. Rather, as will be discussed later, late industrialisation in these cases was heavily sponsored and directed by the state (Chang 2008) in a way extending far beyond that envisaged by Rostow, who saw some of the primary roles of the state as investing in infrastructure, helping diffuse new agricultural technology and managing (increased) foreign aid. For Rostow the state also had a strong role in breaking "traditional" social hierarchies and practices. For him "contact" with Western powers was a positive thing, and development happened through diffusion.

Another stream of modernisation theory had to do with political development. If modernisation theorists saw a relatively strong role for the state in promoting the adoption of modern technology, for example, what type of state would do that? In sometimes strikingly ethnocentric and essentialising (overgeneralise and mischaracterise groups of people) or some might say racist and sexist terms the well-known Harvard political scientist Samuel Huntingdon (1971, 287) identified the issue as the culture and character of people:

> Traditional man is passive and acquiescent; he expects continuity in nature and society and does not believe in the capacity of man to change or to control either. Modern man, in contrast, believes in both the possibility and desirability of change, and has the confidence in the ability of man to control change so as to accomplish his purposes. . . . The differences between a modern polity and a traditional one flow from these more general characteristics of modern and traditional societies.

Of course, this was complete fantasy, even if Huntingdon is still revered by prominent academics such as Steven Radlet (2015) and such tropes are still repeated by leaders such as Nicholas Sarkozy when he was president of France or long-time president of Uganda Yoweri Museveni. On a visit to the continent Sarkozy said, "The tragedy of Africa is that the African has not fully entered into history" even if "modern man" could learn from "African man who has lived in harmony with nature for millennia" (quoted in Ba 2007; Death 2016, 19). Many people considered this to be an extremely racist remark – harkening back to colonial stereotypes about "backward" Africans. While I was a student at the University of Minnesota, I attended a talk by Museveni where he explained

African underdevelopment in relation to a natural abundance of the continent, which meant people didn't have to work as hard as they did in Europe, he claimed.

"Traditional" societies were heavily transformed by the colonial encounter, even if this took the form of "the invention of tradition" (Hobsbawm and Ranger 1983), such as "tribes" which did not exist before European rule. There were certainly ethnic groups before colonialism in the Global South, but ascription of "tribal" identities was a way to "divide and rule", elaborated on later. Such constructions of tradition as backward harken back to the European Enlightenment, which was an intellectual movement which foregrounded reason as the source of authority.

Divide and rule was a well-known Roman strategy in conquered territories and was applied during European overseas colonialism as well. For example, there were no "tribes" in Africa before the Europeans arrived. This was a European concept which they brought with them and imposed in their colonies – sometimes inventing "tribes" for people to belong to, a process known as tribalisation (Ranger 1994). "The chief of a little-known group in Zambia once ventured to remark: 'My people were not Soli until 1937 when the Bwana D.C. [District Commissioner] told us we were'" (quoted in Meredith 2005, 155). The name may derive from the Latin word for soil, which is *soli*. The legal principle of *jus soli* grants citizenship based on birth in a particular territory – "right of the soil". Likewise, the Kalenjin ethnic group in Kenya

> is a recent construct, dating from the mid-twentieth century, when it came to embrace a number of subgroups administered as separate 'tribes' by the colonial authorities. . . . The term Kalenjin literally means 'I say to you' – a direct reference to the linguistic similarity of its members – although significant differences of dialect lead to talk of Kalenjin language clusters.
>
> (Lynch 2011, 3–4)

The development of Kalenjin ethnic group consciousness was partly spurred by a popular radio programme in the 1950s which began with the words "Kalenjin, Kalenjin" (Drogus and Orvis 2012), although sub-ethnic consciousness also continues to remain strong (Lynch 2011). In another example, modern Hinduism in India emerged out of the colonial encounter (Pennington 2005) with input from the British, in addition to serving as an anti-colonial rallying point of identity (Stuenkel 2016).

It could also be argued that the Hutu and Tutsi ethnicities in Rwanda were invented by the German and later Belgian colonists. Prior to colonialism, the distinction between Hutu and Tutsi was a class one, as the latter derived more of their livelihood from cattle ownership, whereas the former were largely peasant farmers. However, colonialism resulted in ethnicisation, as identity cards were introduced and preferential treatment given to Tutsis by the Belgians. The British also appointed or licenced "chiefs" in South Africa – so-called warrant chiefs – and transformed pre-existing institutions. Under colonialism rural areas were marked by a "colonially enforced customary order, which entailed the fusion 'in a single person [of] all moments of power: judicial, legislative, executive and administrative'" (Mamdani 1996, 23 cited in Engberg-Pedersen 2002, 163). Thus, the Global South has been largely produced through, rather than in opposition to, modernity. Indeed, as will be discussed in more detail later, it could be argued that modernisation is now being increasingly driven by emerging powers in the Global South (Gonzalez-Vicente 2017).

Modernisation theory has been and remains influential. Indeed, as will be discussed later, we are arguably seeing a revival of some of its core ideas in current development policies and practices. Its proponents saw it as having a humanitarian dimension – the upliftment of humanity through technological, infrastructural and political development, for example. In practice, however, modernisation theory as implemented through development policy was infused with dramatically unequal power relations and a strong geopolitical impetus. This was to feed into later critiques, some of which I now turn to.

Further reading

Books

Apter, D. 1965. *The Politics of Modernization*. Chicago, IL: University of Chicago Press.

Rostow, W.W. 1960. *The Stages of Economic Growth: A Non-Communist Manifesto*. Cambridge: Cambridge University Press.

Articles

Abumere, S.I. 1981. "The Geography of Modernisation: Some Unresolved Issues." *GeoJournal* 5 (1): 67–76.

Huntingdon, S. 1971. "The Change to Change: Modernization, Development, Politics." *Comparative Politics* 3 (3): 283–322.

Klinghoffer, A. 1973. "Modernisation and Political Development in Africa." *The Journal of Modern African Studies* 11 (1): 1–19.

References

Amsden, A. 1989. *Asia's Next Giant: South Korea and Late Industrialization*. New York and Oxford: Oxford University Press.

Anyangwe, E. 2017. "Brand New Macron, Same Old Colonialism." *The Guardian*. www.theguardian.com/commentisfree/2017/jul/11/slur-africans-macron-radical-pretence-over.

Ba, D. 2007. *Africans Still Seething Over Sarkozy Speech*. https://uk.reuters.com/article/uk-africa-sarkozy/africans-still-seething-over-sarkozy-speech-idUKL0513034620070905.

Chakrabarty, D. 2000. *Provincializing Europe: Postcolonial Thought and Historical Difference*. Princeton, NJ and Oxford: Princeton University Press.

Chang, H. 2008. *Bad Samaritans: The Guilty Secrets of Rich Nations and the Threat to Global Prosperity*. London: Random House Business.

Chipungu, S. 1992. "Accumulation from Within: The Boma Class and the Native Treasury in Colonial Zambia." In *Guardians in Their Time: Experiences of Zambians Under Colonial Rule, 1890–1964*, ed. S. Chipungu. London: Palgrave Macmillan.

Cox, R.W. 1987. *Production, Power and World Order: Social Forces in the Making of History*. New York and Guildford, Surrey: Columbia University Press.

Death, C. 2016. *The Green State in Africa*. New Haven and London: Yale University Press.

Derickson, K.D. 2015. "Urban Geography I: Locating Urban Theory in the 'Urban Age'." *Progress in Human Geography* 39 (5): 647–57.

Drogus, C., and S. Orvis. 2012. *Introducing Comparative Politics: Concepts and Cases in Context.* 2nd ed. Washington, D.C.: CQ; London: SAGE [distributor].
Engberg-Pedersen, L. 2002. "The Limitations of Political Space in Burkina Faso: Local Organizations, Decentralization and Poverty Reduction." In *In the Name of the Poor: Contesting Political Space for Poverty Reduction*, eds. N. Webster and L. Engberg-Pedersen. London and New York: Zed Books; New York: Distributed by Palgrave Macmillan.
Gonzalez-Vicente, R. 2017. "South – South Relations Under World Market Capitalism: The State and the Elusive Promise of National Development in the China – Ecuador Resource-development Nexus." *Review of International Political Economy* 24 (5): 881–903.
Hall, D. 2013. *Land.* Cambridge: Polity.
Hobsbawm, E., and T. Ranger, (eds.) 1983. *The Invention of Tradition.* Cambridge and New York: Cambridge University Press.
Huntingdon, S. 1971. "The Change to Change: Modernization, Development, Politics." *Comparative Politics* 3 (3): 283–322.
Khoo, S. 2013. "Sustainable Development of What: Contesting Global Development Concepts and Measures." In *Methods of Sustainability Research in the Social Sciences*, eds. F. Fahy and H. Rau. Thousand Oaks, CA: Sage.
Korten, D. 1999. *The Post-Corporate World: Life after Capitalism.* San Francisco, CA [United Kingdom]: Berrett-Koehler.
Lee, M. 2014. *Africa's World Trade: Informal Economies and Globalization from Below.* London: Zed Books.
Lynch, G. 2011. *I Say to You: Ethnic Politics and the Kalenjin in Kenya.* Chicago, IL: University of Chicago Press; Bristol: University Presses Marketing [distributor].
Mamdani, M. 1996. *Citizen and Subject: Contemporary Africa and the Legacy of Late Colonialism.* Princeton, NJ and Chichester: Princeton University Press.
Meredith, M. 2005. *The State of Africa: A History of Fifty Years of Independence.* London: Free Press.
Myers, G. 2011. *African Cities: Alternative Visions of Urban Theory and Practice.* London: Zed Books.
Myers, G. 2018. "The Africa Problem in Global Urban Theory: Reconceptualising Planetary Urbanization." *International Development Planning Review* 10: 231–253.
Nally, D. 2011. *Human Encumbrances: Political Violence and the Great Irish Famine.* Notre Dame, IN: University of Notre Dame Press.
Pempel, T.J. 1999. "The Developmental Regime in a Changing World." In *The Developmental State: Ithaca*, ed. M. Woo-Cummings. New York: Cornell University Press.
Pennington, B. 2005. *Was Hinduism Invented: Britons, Indians, and the Colonial Construction of Religion.* New York and Oxford: Oxford University Press.
Pierce, J., D.G. Martin, and J.T. Murphy. 2011. "Relational Place-Making: The Networked Politics of Place." *Transactions of the Institute of British Geographers* 36 (1): 54–70.
Radlet, S. 2015. *The Great Surge: The Ascent of the Developing World.* New York: Simon and Schuster.
Ranger, T. 1994. "The Tribalisation of Africa and the Retribalisation of Europe." In *St Antony's Seminar Series: Tribe, State, Nation.* Oxford.
Rostow, W.W. 1960. *The Stages of Economic Growth: A Non-Communist Manifesto.* Cambridge: Cambridge University Press.
Selwyn, B. 2014. *The Global Development Crisis.* Cambridge: Polity.
Shubin, V. 2008. *The Hot 'Cold War': The USSR in Southern Africa.* London: Pluto.
Stuenkel, O. 2016. *Post-Western World: How Emerging Powers Are Remaking Global Order.* Cambridge: Polity.
Truman, H. 1949. *Inaugural Address.* https://www.trumanlibrary.org/whistlestop/50yr_archive/inaugural20jan1949.htm
Williams, E. 1994. *Capitalism and Slavery.* Chapel Hill: University of North Carolina Press.
Wright, D. 2010. *The World and a Very Small Place in Africa: A History of Globalization in Niumi, the Gambia.* 3rd edition. Armonk, NY and London: M.E. Sharpe.

CHAPTER

2

Early critiques

Dependency, world systems and alternative theories of development

According to modernisation theorists, as long as the pre-conditions were in place and people were willing to accept new cultural practices and shed the old ones, then the whole world could follow in the footsteps of the United States and Western Europe and achieve high mass-consumption societies – "the American Dream". However, the continuing, and sometimes deepening, reality of mass poverty in much of the world led to the development of alternative theories: in particular, the schools of structuralist economics, dependency and world systems theory.

Structuralist economics

One of the most influential centres of new thinking on development and development economics has been the United Nations Economic Commission for Latin America and the Caribbean (ECLAC), also known by its Spanish abbreviation, CEPAL (Interestingly, many Western European countries, the US, Canada, Japan and South Korea are members of this too, bespeaking their global economic and political influence). Probably the most influential figure historically in CEPAL was Raúl Prebisch, who became its director in 1950. He is noted for two related and major innovations in development theory. The first was the idea that rather than the global market economy leading to economic convergence between world regions, it tended to be regionally polarising, creating rich "cores" and poor "peripheries", which was prefigured by Harold Innes's ideas of heartland and hinterland in the 1930s (Siakwah 2017). Prebisch identified trade as a primary mechanism through which this outcome developed.

The founder of conventional trade theory, David Ricardo, had posited that free trade would lead to mutually beneficial economic development between partners. He gave the example of the trade in wine and cloth between Portugal and England, arguing that even if Portugal could produce both wine and cloth cheaper than England, it would benefit its economy most to specialise in that economic activity in which it was most efficient, that is, the activity in which it had what he called a comparative advantage. People sometimes refer to Ricardian economics as that school of economics that promotes so-called "free trade".

His argument makes intuitive sense. If Portugal was more efficient (could generate higher profits) in producing wine, it would make sense for it to devote its land, labour and capital, or what economist call factors of production, to that and to import cloth from England, the argument went. Likewise, even if English cloth was initially more expensive than Portuguese cloth, as capital and labour moved into producing wine, that would open up space for England to begin supplying cloth to Portugal. However, in order for this model to work, all sorts of untenable assumptions have to be made. Things such as there being no economies of scale. But economies of scale, where bigger factories are more efficient and consequently can outcompete small ones, are rife in manufacturing. By way of simple illustration, there is something in the chemical industry called the two-thirds or 0.6 rule. The capacity of chemical vats increases disproportionately to the amount of steel used to make them. A 0.6 increase in the amount of steel used results in a doubling of volume of a vat, meaning that larger plants can be more efficient as they use fewer inputs (Kaplinsky and Cooper 1989).

There are also major differences in infrastructure and skill levels across places, so "first movers" in terms of industrialisation tend to have a whole set of advantages over "late-comers", even if labour costs are lower in the latter, which the famous economic historian Alexander Gershenkron argued was one of the "advantages of backwardness". Generally, however, the "advantages of backwardness" can only be effectively leveraged through an effective industrial policy, which protects late-comer industry from the higher productivity of industries located elsewhere and encourages and guides their development in other ways, discussed in more detail later.

As noted earlier, the primary mechanism that Prebisch identified through which some places developed and others were underdeveloped was through the terms of trade. This was Prebisch's second major innovation, and it became known as the Prebisch-Singer thesis, as Hans Singer of the Institute of Development Studies in Brighton in the UK initially elaborated it. The basic principle is quite simple. As peoples' incomes rose in richer countries during the post–World War II boom in Europe and North America, for example, they tended to spend extra income on manufactured goods and services and not proportionately as much on additional food, which might be imported from the tropics. The observation that as people's income rises the total proportion spent on food tends to fall was first made by the German statistician Ernst Engel in the nineteenth century and is known as Engel's law. So what economists call the income elasticity of demand tends to be higher for manufactured goods through time – that is, as people's incomes go up, they tend to demand relatively more manufactured products, such as smart phones or Fitbits, for example. Given the law of supply and demand, what this means is that prices for primary commodities (such as minerals or food) tend to go down through time relatively compared to those of manufactured goods. The effect of these tendencies, Prebisch and Singer argued, was to lock primary commodity producers into unfavourable terms of trade (relative prices for their imports and exports) and to make them poorer through time as the prices for their exports declined and the prices for their imports (often mostly manufactured goods) increased, which will be discussed in more detail in the next chapter when talking about the impacts of

World Bank/IMF programmes. The long-term or secular trend of declining commodity prices continues, although demand from China in particular was associated with a recent reversal in the terms of trade (see Figure 2.1). However, as the graph shows, primary commodity prices are notoriously volatile, making planning and economic conditions unstable in primary commodity exporters.

European colonialism in the Global South had sought to make it a commodity exporter and importer of manufactured goods. The policy prescription which was to flow from the Prebisch-Singer thesis was that countries of the Global South should try and break out of this "trade trap" by industrialising. The fact that the already developed countries had established competitive advantages made this exceptionally difficult under conditions of "free trade" though. The route to industrialisation was then seen to be through so-called import-substitution industrialisation (ISI).

ISI was widely adopted by countries in the Global South after the Second World War and can be considered a form of nationalist-oriented modernisation. The basic theory behind ISI was quite simple: that in order to develop, countries needed to industrialise. It was argued that because the already rich countries were more competitive in manufacturing than the poorer countries of the world, in order for them to industrialise, they had to shelter their industries from international competition in order to give them the breathing space to develop, to learn how to produce manufactured goods and capture the benefits of the domestic market, which was still often being serviced by companies from former colonial powers, largely through imports. This is known as "infant industry" protection, the idea being that as manufacturing industries were built up and became competitive, they could begin to export.

Foreign exchange (forex) or currency is a relatively scarce resource in low-income countries; although, of course, this is at a national level and varies substantially within societies depending on class positions. As the prices for raw materials or primary products have a long-term or secular tendency to go down, this tends to make it even scarcer for many countries in the Global South dependent on them for their exports. ISI was meant to have the added benefit that it would conserve scarce foreign exchange, as countries would now produce their own manufactured goods, rather than having to spend forex on importing them. However, in practice ISI tended to be forex-intensive. There were a number of reasons for this. First, setting up new industries requires imports of what economists call capital goods – the machines that make or help make things. So, for example, sewing machines are capital goods if they are used to produce garments for sale. In the textile industry bigger machine tools would include things like those for spinning, which turn cotton into thread, which can then be woven into fabric by weaving machines (see Figure 2.2 for an example of spinning machines).

The more technologically sophisticated the industry is, the more expensive the machine tools associated with it tend to be. The textile and clothing industry is often thought to be the easiest industry to begin the industrialisation process with because the skills needed to make clothes are relatively easy to master and it is labour, rather than capital, intensive – that is, it requires relatively large amounts of labour, as compared to capital or machinery and buildings. Nonetheless, sewing machines and other equipment often have to be imported to begin the process of setting up the industry, costing forex.

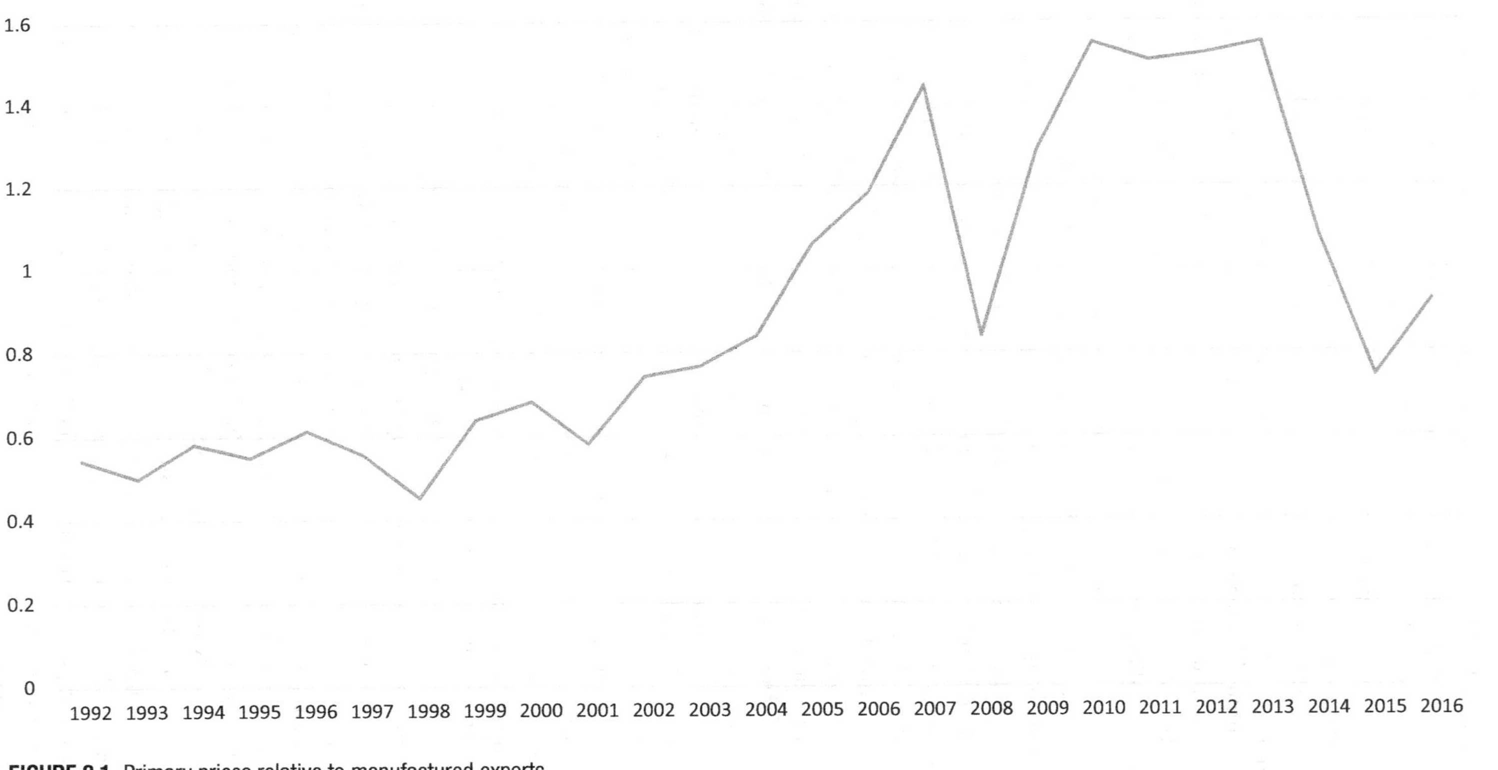

FIGURE 2.1 Primary prices relative to manufactured exports

Source: Calculated from the IMF (2018) and World Bank Manufacturing Unit Value Export Index (n.d.)

FIGURE 2.2 Spinning machines

If infant industries are protected by tariffs or other restrictions on imports, it is typically because they are not internationally competitive. If they are not internationally competitive, they won't export and earn foreign exchange, so even though the initial theory was that ISI would save forex, in practice, this often turned out not to be the case. Even if fewer clothes were imported, almost by definition, unless they were to be subsidised by the government, they would be more expensive than imports, and the machines to produce them would cost scarce foreign exchange to import, often financed by overseas loans. Because many of the countries that adopted this strategy were poor, the home market also turned out to be quite limited. Small domestic markets meant that economies of scale in many industries were not realised, making these industries relatively less efficient compared to overseas competitors, although spurring rapid growth in some countries in Latin America, for example, for certain periods of time, as under Juan Perón in post–Second World War Argentina (the foundational importance of resources is revealed in the county's name, from the word for silver). However, there were also distributional impacts of this policy choice.

For the sake of argument, let's stick with the example of clothes. In many low-income countries, and even some middle-income countries such as China, the majority or much of the population is rural (around 40 per cent in China's case). If domestically produced clothes are more expensive than imported ones and the majority of the population is rural, this represents an effective tax on them, some would argue. While some urban workers would benefit from the jobs created in the clothing industry, rural people have to pay more for their clothes, thereby reducing their incomes. In the 1970s Michael Lipton (1977) argued that polices such as these, and taxes levied on farmers across much of the developing world to fund ISI, represented what he called "urban bias", where city populations often got subsidised food and other benefits at the expense of the more numerous rural population. It could, however, be counter-argued though that the long-term benefits of industrialisation were such that the initial costs were worth it. A potential problem with this approach might be that as the famous British economist, John Maynard Keynes, put it "we are all dead in the long term", meaning that such a strategy, with promised success in the future, might be cold comfort to farmers, for example, who saw their incomes reduce in the present. However, in many cases ISI did not lead to productivity growth or getting more output per unit of input, such as labour, essential to, or some might say the definition of, industrialisation.

As noted earlier the theory was that after an initial period of being sheltered, manufacturing industries would mature and through time become competitive. However, a problem with many instances of ISI was that the "infant industries" never grew up or matured, as it was possible for them to be both profitable and relatively inefficient, with largely captive domestic markets, while still often continuing to be foreign exchange–intensive. There were, however, quite varied experienced with ISI, and it is often a necessary step in the process of industrialisation (for a discussion of the benefits of ISI see Peet and Hartwick 2009). In some cases, such as South Korea, it was successful when combined with export promotion to earn foreign exchange and widen markets. Successfully developed countries, such as the United States, often industrialised on the basis of ISI, even if it wasn't called that at the time, sometimes because they already had relatively large domestic markets, meaning there was still intense competition between firms, which could spur productivity growth. Also some argue that ISI was stopped prematurely across much of Africa, for example, before it had the time to yield expected dividends (Riddell 1990).

Thus the way in which ISI was implemented was important, as were the initial contexts. For example, during the period of the "Brazilian miracle", when economic growth was sometimes close to 10 per cent from the late 1960s to the early 1980s, ISI was implemented by the military dictatorship. However, the structure of industry created during this time was problematic. For example, in the auto industry many of the cars built were made by TNCs and were merely assembled from what are called totally knocked-down-kits imported from abroad to avoid tariffs on assembled cars. This strategy is known as tariff jumping but meant that there was little technological development taking place in Brazil when this happened, and there were very limited linkage effects to the local economy in terms of things like raw materials and components, even if some assembly jobs were created. Limited linkage effects also translate into fewer of what economists call multiplier effects.

A multiplier effect arises when a unit of currency, such as a dollar, gets spent multiple times in the same economy, generating additional economic activity. So if steel was being sourced nationally for the car industry, for example, more steel workers would be employed and profits would be made by the steel mill owner. These people would then spend their money buying things in the local economy, and that would generate demand for other industries, such as food or clothing. Alain De Janvry (1981) distinguished between what he called export-oriented disarticulated economies and import-substituting ones. In what was then an articulated economy demand from capital goods, intermediate goods (like steel) and consumer goods industries reinforced each other. In disarticulated economies there weren't the same kind of linkages for the reasons outlined earlier, in the case of Brazil, or for export-oriented ones, such as many African economies dependent on primary commodity exports.

Dependency and world systems theory

Whereas structuralist economists envisage a continuation of the capitalist economic system in the Global South through its transformation via industrialisation, other critiques argued that there was a need to either detach from capitalist economic logics or move towards socialism. The two most well-known of these theories are dependency and world systems theory. As noted earlier, dependency theory largely originated in Latin America, and to a lesser extent Africa, although it was also elaborated by Western theorists such as Frank (1967) and Evans (1979), though using Latin American examples. As noted earlier, Cardoso and Faletto were some of the most important initial dependency theorists. They argued that capitalist development produced nested hierarchies of relatively richer cores and poorer peripheries. At a global level Europe and North America were the core and the countries of the Global South the peripheries. However, there are also cores and peripheries within countries. Rio de Janeiro and Sao Paulo would be some of the core regions in Brazil and the poorer northeast of the country some of the periphery, for example.

World systems theory basically developed out of, or follows on from, dependency theory. It is most often associated with Immanuel Wallerstein (1975). The key distinction between dependency and world systems theory is that whereas many early dependency theorists saw peripheries being condemned to permanent marginality and exploitation unless they delinked from the global system and established their own, more independent economic developmental dynamics, world systems theory allowed for mobility within the international hierarchy. Some later dependency theorists did, however, allow for the possibility of what they called "dependent development", where average incomes might rise but at the cost of extreme inequality and dependence on other countries. Interestingly, one of this later group, Fernando Cardoso, later went on to become president of Brazil after introducing several neoliberal policies previously as minister of finance.

Rather than seeing the world as being divided into two mega-regions of core and periphery, Wallerstein also said there was a third region: the semi-periphery – a kind of intermediate zone, where the richer countries of Latin America might sit, for example. Mobility between the zones was possible, he argued, so peripheral countries might become semi-peripheral, for example, depending on developments in the global economy and their own internal dynamics.

Neo-Marxist theory and imperialism

Depending on the author, dependency and world systems theory could be considered Neo-Marxist, in that they argue that capitalist development in the Global South often leads to underdevelopment, rather than development. However, some draw a distinction between these approaches and Neo-Marxism because of the emphasis on trade and exchange relations, rather than relations of production and class. Of course, it is important to examine both, and trade and class relations are partly created through interaction with each other or are inter-constitutive. Class refers to patterned social differentiation between groups based on income and wealth. Weberian definitions emphasise income and occupational differences, whereas Marxist definitions focus on who owns the means of production (capital), such as factories and offices, and who then have to work for these people (labour).

While a previous Catholic pope, Benedict, praised Marx's "great analytical skill", there are sometimes somewhat contradictory passages in Marx's writing. At certain points he talks about more advanced countries merely showing the less advanced the vision of their future. While he deplored the brutality of colonial conquest, he saw capitalism as a progressive force as it developed the productive forces of society, which could provide the material means to move towards socialist and later communist prosperity. So-called orthodox Marxists see capitalism as a progressive social force in the Global South. The most well-known of these was Bill Warren (1980) and also Sender and Smith (1986).

In other passages of his writing Marx was less sanguine about the impacts of capitalism in the Global South. Neo-Marxists argue that global capitalism results in underdevelopment in the Global South. Whereas some dependency theorists were content to recommend delinking rather than the abolition of capitalism, Neo-Marxists tend to recommend moving towards socialism, as they view capitalism as a retrogressive and destructive force, in particular, as it manifests through the phenomenon of imperialism. What is the nature of contemporary imperialism?

While the era of formal colonialism is over, except for those indigenous people living in settler colonies, the economic imperatives which gave rise to it still exist – namely the desire to open up new markets, sources of raw materials, access to cheap labour power and other ancillary motives around ensuring geopolitical power and security. These imperatives generate incentives for the major industrial powers to design geogovernance (power projection across borders) arrangements that achieve largely similar economic results to colonialism – resource extraction and market access – without the administrative burden and resistance associated with direct territorial control.

As their economies have become more inter-woven, inter-connected and interdependent through trade and financial flows, the incentives to go to war to assert territorial control by the major industrial powers has also diminished. Such direct territorial control is, in any event, no longer necessary, as the relations of politico-economic dependency established under colonialism have remained largely intact and remain functional and effective in surplus extraction from the periphery to the core of the global system. This then raises the question of what types of governance arrangements are currently in place to maintain and oftentimes deepen uneven development and inequality globally.

There is an ongoing debate in the literature as to whether or not the concept of imperialism is still a useful one (see Kiely 2010 for a review). Marx (1894 [2011]) in *Das Capital* (Volume 3) identified a tendency for the rate of profit to fall in capitalism.

Later Lenin (1917 [2010]) argued that the shift from competitive to monopoly capitalism, where there were large industrial conglomerates, was responsible for imperialism. What is imperialism, and does it still exist, or are there other concepts which are more appropriate to examine the current conjuncture?

Under colonialism, imperialism could be understood as the penetration of capital into pre-capitalist areas accompanied by the use of political force. This definition shares similarities with Marx's idea of primary or primitive accumulation, recently reformulated by David Harvey (2003) as 'accumulation by dispossession'. According to Marx, primary or primitive accumulation is the way in which capital is originally generated. Marx views capital as a social relation, where some people own the means of production (capitalist class) and others have to work for them (labour). Primary accumulation refers to practices, such as the enclosure of the commons in England where people were forcibly displaced from land so that it could be privatised by the emergent gentry class.

Other definitions of imperialism might include the domination of the people of one nation-state by another through 'rule at a distance' (Abrahamsen 2000). However, there has been substantial debate about whether the focus on the nation-state is still an appropriate one. According to Kemp (1972, 29) "American imperialism rules over 'an empire without frontiers' which has no parallel in the past". Indeed, the American essayist Gore Vidal argued the American empire was one of the most successful in history because no one knew it was there (Sullivan 2001). However, others have argued that with the increasing deregulation of the global economy, transnational capital now exercises sovereignty, or ultimate decision-making power, going so far as to suggest that we should now talk of the 'state form of capital' rather than states as distinct social forces (Woodley 2015). However, adopting a 'strategic-relational' approach (Jessop 2002) to the state would complicate this. In much social science literature the state is either seen as an actor in its own right or as something which is controlled by other social forces. Karl Marx famously wrote that the state was no more than an "executive committee" of the bourgeoisie, or capital. A strategic-relational approach, however, recognises that the state is both a social relation embedded in the wider society and in some sense an actor, whose decisions have strategic impetus and force.

According to Kiely (2010) the primary way in which imperialism is expressed in the contemporary period is through restriction on the adoption of industrial policies by developing countries, whereby they would seek to shelter and develop their own companies so that they might, in time, be able to compete with those of the more developed countries, and also now NICs, such as China. This imperialism can be achieved because it is the major industrial powers which control the majority of votes at the international financial institutions (IFIs) of the World Bank and International Monetary Fund (IMF), with the United States still being the only country to hold sufficient votes to have veto power in both institutions.

Power inequality is central to the operation of the IFIs, even if the strap line for the World Bank was that it dreamt of a world free of poverty. If poverty is not conceptualised as an outcome of power inequality, this is likely to remain a dream. I undertook a World Bank consultancy on teacher education in Malawi in 2007. During this project I worked with highly dedicated and knowledgeable people. However, in the office, power dynamics between some World Bank staff and government officials were very evident. I particularly remember one staffer, who wasn't involved in our project, on the phone telling an administrator when they wanted to see a government minister.

The neo-imperial imperative

There is substantial literature on the motivations behind colonialism. While there are a variety of theories and a mix of motivations, it is now generally accepted that the primary motivations were economic. In relation to industrial capitalist colonisation, the scramble for Africa of the late nineteenth century was set off by the need to find new markets for the products of European industry, given depressed economic conditions on the continent (Pakenham 1992) and the need to find new sources of raw materials, particularly in the context of the global 'cotton famine' or shortage which resulted from the American Civil War (Mamdani 1996). Some might draw a distinction between the imperial era, during formal colonialism, and imperialism as a stage of capitalist development where value is extracted from overseas territories through foreign investment. I do not draw that distinction here, as the motivations for power projection across borders or territories, in particular, around access to markets, raw materials and geo-strategic influence, remained largely unchanged from the colonial period. As John Goodman's character notes in the British Broadcasting Corporation series *Black Earth Rising*, they may not be wearing pith helmets anymore but they are still playing the "Great Game", which referred specifically to geopolitical competition between Britain and Russia in the nineteenth century, but is used more generally to refer to "great" power territorial competition. Although in fact Melania Trump, the US First Lady, controversially did wear a pith helmet, associated with colonial rule on the continent, when she was on her African tour in 2018.

There has been a round of global economic restructuring associated with neoliberalisation. As the post-war economic boom came to an end and rates of profit fell in the North Atlantic economies in the late 1960s and early 1970s, there was again a shift from geographically intensive, or more autocentric dynamics, to a geographically extensive regime of accumulation associated with intensified globalisation (Amin 1994; Aglietta and Fernbach 1979). What does this mean? Wages are both a cost and benefit for capitalists. In general, to maximise profit it is best for capitalists if they pay their own workers low wages but for other firms to pay their workers higher wages to create a market for the goods they want to sell.

Capitalists understand that workers' wages are, or can be, depending on the industry, an important source of demand for their products. The American auto entrepreneur Henry Ford famously paid his workers five dollars a day so that they could afford to buy his cars. In the post-war era this idea was extended to national economies in the North Atlantic, as rising wages, negotiated between big business, big labour (strong trade unions) and the state, created a virtuous circle of rising wages and productivity, which also increased profits. However, with the advent of neoliberalism and the shift towards free trade, the logic of wages as a cost came to prominence in the North Atlantic economies, and capitalists, again depending on sector and country, are incentivised to try to cut wages or keep them low to become or remain competitive. Although there is also something called efficiency wages. If workers are underpaid, they may be demotivated and less productive, so it may, in certain circumstances, benefit capitalists to pay higher wages to make their workers more productive. The Reagan and Thatcher governments, however, sought to revive rates of profit in the US and UK economies for companies and corporations through 1) disciplining/eviscerating the labour movement domestically and internationally and 2) opening up markets around the world for exports – the main type of imperialism today, according to Ray Kiely.

While the United States still has by far the world's biggest economy, measured at nominal prices (the value of economic output in dollars, rather than adjusting them relative to what they can buy), and is still by far the world's most influential state, its (always incomplete) "hegemony" in the international system is under threat in the last decade and a half in particular. Writing in the early 1990s Robert Jackson claimed that "the West is now more secure and confident in the superiority of its values than it has been at any time since the end of the Second World War" (Jackson 1992, 13). However, the rise of Islamic fundamentalist terrorist groups around the world and potential state challengers, particularly China, and the industrial hollowing out of the American economy and stagnating real wages have recently undermined this confidence. For example, a recent poll found that only 28 per cent of Americans agree with the statement that the US "stands above all other countries in the world" (Pew Research Centre 2014). This partly explains why 2016 presidential candidate Donald Trump's message to "Make America Great Again" resonated with a substantial portion of the population. Nonetheless, the United States and its Western allies continue to exercise very substantial influence bilaterally and collectively through the institutions they largely control, such as the dominant IFIs and the North Atlantic Treaty Organization (NATO), and through more informal mechanisms of coordination such as the Group of 7 (G7) and the World Economic Forum (WEF). The fact that the G7 is more than a "talking shop" is evidenced by the fact that it has recently signed an agreement with a number of African countries around agricultural "development" based on privatisation of land and liberalisation. This programme has recently been heavily criticised by members of the European Parliament (Guardian 2016). However, according to the late Samir Amin (2016) NATO is the key structure of Western power through its recourse to the use of force.

While under international law the United Nations is meant to be the body which has a monopoly on the legitimate use of force internationally, unless states are attacked or an attack is in preparation, NATO has at times arrogated this power onto itself, beginning with the bombing of Yugoslavia in the late 1990s, which was conducted without authorisation under the Charter of the United Nations (Mandel 2004). Shaw (2003) talks of an emergent global state with a Western core. However, this is arguably a global capitalist "state", which combines the promotion of global accumulation with legitimation through aid programmes, support for NGOs and, where its military or sometimes economic security is threatened or challenged, the use of force.

How Western-led economic restructuring and geopolitical power are expressed or play out in particular places in the Global South varies depending on factors such as the nature of their states and their geo-strategic positionality. For example, as will be discussed in more detail later, the fact that Ethiopia is geo-strategically important for the Western powers in their so-called "war on terror" has meant they have given the government more aid and greater latitude to develop its own economic policies than many other developing countries. The phenomenon of offshoring of labour-intensive elements of manufacturing value chains has benefitted some countries, primarily in Asia, but generally not other regions of the Global South in Africa or Latin America. Thus, the relationship between imperialism and development is a complex one. While the well-known geographer David Harvey recently courted controversy by arguing that under globalisation the complexity of flows meant that imperialism has faded or been reversed, with the rise of Asia, in particular, others dispute this (Smith 2018).

If imperialism is taken to mean the domination of one state or group of states over another state or group of states, then its impact on development is indeterminate. It is still possible for a country to be dominated by another state or group of states and to experience (dependent) development. Whether or not this is the case partly depends on what some have called territorial versus "non-territorial" forms of power. Territorial logics of power relate to the state. For example, during the Cold War the US was willing to tolerate, and even subsidise, capitalist competitors such as South Korea in order to ward off the spread of communism. Thus the territorial logic superseded the "non-territorial" logic, which would have been around opening up South Korea's market to American corporations' exports and investment, for example. The well-known Canadian-based political scientist Stephen Gill argues that even though the European Union (EU) has a bigger economy than the United States, the US is still the pre-eminent power because it is able to dominate in its bilateral relations with each of the individual states that comprise the EU. The developmental impacts of imperialism then vary from context to context, even if the overall relations continue to be unbalanced and structured by extreme power inequality.

Socialist development?

The logical outcome of these types of more critical theories of development described earlier is that there should be a move away from capitalist development towards socialism. There have been a variety of different approaches to this in the post-colonial period, most of which have failed to deliver anticipated benefits and/or have been highly authoritarian and have collapsed.

Pan-Africanism is the idea that Africans and the African diaspora should strengthen their bonds and solidarity in order to more effectively counter marginalisation and racism. In practical terms it sometimes finds expression in the idea that there should be a "United States of Africa", for example. One of the "fathers" of Pan-Africanism, the first president of independent Ghana, Kwame Nkrumah, wrote about the need to transcend what he termed neo-colonialism, or what the first president of Tanzania, Julius Nyerere, called "flag independence" – which is the idea that even though a country may be nominally independent, its economic dependence gives foreign powers and companies a determining say in how it is run. However, levels of autonomy from the Western powers vary dramatically across the Global South.

In some cases Western powers have supported the overthrow of governments they saw as hostile or inimical to their interests, such as the Mossadegh government in Iran in the 1950s or more recently the government of Saddam Hussein in Iraq in 2003. In other cases governments in the Global South have successful resisted Western intervention or interference, with China being perhaps the prime example of that, with its economic reform programme initiated under Deng Xiaoping in the 1970s dramatically and successfully transforming that country's economic structure, first through the liberalisation of agricultural marketing, so farmers could keep some of the profits from the sale of their crops, the development of township and village enterprises and the setting up of special economic zones in the south of the country to attract foreign investment to produce for export.

The scale of economic transformation wrought in China in the space of a few decades is vast, with hundreds of millions of people lifted out of poverty. Flying into Shenzhen, which is over the (internal) border from Hong Kong, over the Pearl River Delta provides a breathtaking view of the scale of development. The current Chinese president, Xi, talks of "socialism with Chinese characteristics", which in practice means further economic liberalisation and the deepening of capitalism in that country, with some exceptions, such as the reinstitution of large-scale public housing projects in 2009 (Chen *et al.* 2013). China's transition away from socialism as a system of economic organisation has been more gradual than in many other cases.

According to neo-conservative theorists such as Francis Fukuyama (1992), free-market capitalism and liberal democracy represent the apogee, or highest form, of human development. Private property is meant to offer protection from the state, and the separation of powers between legislative (parliament), judicial (legal system) and executive branches of government are meant to prevent overweaning government and authoritarianism. Where firms, and consequently markets, are highly developed, this arguably is the case. Socialism has historically, however, designated a very strong role for the state in planning or attempting to plan economic activity and consequently has tended to degenerate into authoritarianism, even if this was not the initial intention.

Socialist economies in general tend to suffer from what János Kornai (1992) called the "soft budget constraint". In a capitalist economy, if a firm does not use resources efficiently, it gets outcompeted by other firms and goes out of business. However, when governments run productive enterprises, they are not constrained by profitability and there is often a reluctance to shut down loss-making enterprises because this would generate unemployment, reduce demand in the economy and restrict the supply of goods, thereby potentially compromising regime legitimacy. This is not to say that state-run enterprises can't be efficiently run – they can. Just look at the role which state-owned enterprises (SOEs), such as the Industrial and Commercial Bank of China or State Grid, which are some of the biggest companies in the world, have played in China's economic transformation. Likewise the Pohang Steel Corporation (POSCO) in South Korea was at one time the world's most efficient steel plant despite/because of the fact it was state-owned (Amsden 1989). There are different definitions of what constitutes a state-owned enterprise, but one is that it is one where the state exercises a significant degree of control by virtue of either complete or partial ownership. In China the state also exercises significant influence over private companies, which are required to have a Communist Party representative if there are more than 50 employees and a cell if there are more than 100.

There are models of commercialisation where SOEs are run on lines of profitability or partial privatisation, which have been extensively pursued around the world, which can also increase efficiency. However, to date, socialist political economies have tended to be inefficient and lacking in innovation, sometimes environmentally destructive and authoritarian as the state assumes such a dominant role in the economy and often forbids the existence of other political parties. In some cases, the results have been genocidal. For example, when the Khmer Rouge came to power in Cambodia in the 1970s it depopulated the cities with devastating economic consequences in addition to killing more than a million people and burying them in the so-called "killing fields". The experience of "actually existing socialism" has been less dramatic and devastating elsewhere but has crumbled almost everywhere except Cuba, North Korea and, to the extent that it has a socialist economy, Venezuela, although it has been associated with economic collapse, with inflation running at more than 1 million per cent in 2018 and rising authoritarianism in that country. The Cuban experience has been more mixed and stable, but nonetheless highly authoritarian.

The Soviet Union used to be Cuba's largest market for its sugar until the Soviet Union's collapse in the early 1990s. However, after its collapse the Cuban government shifted to low-input agriculture as it could not afford to pay for the same amount of imported oil that it had previously, for example. Nonetheless, Cuba is the only country in the world to have a "very high" score on the United Nations' Human Development Index while living within an ecological footprint of 1.7 "global hectares" per person (Global Footprint Network 2015). A global hectare represents the average productivity of a hectare around the world in a year. However, most countries that experimented with it have transitioned in different ways away from socialism.

Case study: African socialism in Tanzania

Tanganyika became independent from Great Britain in 1961 and three years later joined with the island state of Zanzibar to become Tanzania. Until the end of the Second World War, Tanganyika had been a German colony, but became in effect a British one afterwards, mandated by the League of Nations – the historical precursor to the United Nations. Colonial rule in Tanganyika was brutal, as it was in the rest of the continent. For example, there were instances where the British tarred porters' feet as they were wearing through to the bone, rather than giving them shoes, which would have been more expensive (Schraeder 2000). Zanzibar was nominally or notionally independent, but the British insisted on approving who would be sultan. It was the site of the world's shortest war – lasting around 40 minutes – in 1896 when the British imposed their preferred candidate through literal "gunboat diplomacy" by shelling the sultan's fort.

Tanzania's first president was Julius Nyerere – one of the only university graduates in the country at independence, having attended the University of Edinburgh in Britain. He believed in what he called African Socialism. This was the idea that socialism or a more communal society had existed before European colonialism in Africa and that lessons could be learnt from this previous method of social organisation. Nyerere's signature initiative was something called ujamaaisation, or the creation of "ujamaa" villages. Ujamaa roughly translates from Swahili as community.

The theory behind moving rural people to villages was that it would allow their upliftment by enabling them to have access to schools and health services, for example. However, the way in which villagisation was enforced was authoritarian and led to the forced closure of genuinely community-driven projects, such as the Ruvuma Development Association, which had provided inspiration for the programme of ujamaaisation, although this was a decision of the ruling Tanganyika African National Union political party, not Nyerere. Another problem with the villagisation programme was that in what were often relatively fragile agro-ecologies people had developed ways of "living lightly" on the land over millennia. However, concentrating people in villages meant that land surrounding them became more intensively used, leading to environmental degradation.

In terms of industrial strategy, Nyerere's government favoured something called the Basic Industries Strategy, developed by the Caribbean economist Clive Thomas, currently head of the Guyana Sugar Corporation and a presidential advisor on Sustainable Development there. Thomas (1974), who was an economic advisor to the Tanzanian government at the time, argued that there were certain industries that generated more added value than others, which he called basic industries, such as glass and steel – what economists called intermediate goods industries, which are used to make final goods, such as glasses. The prescription from Thomas's work was that Tanzania should begin to build up capacity in these and other industries which could generate exports and foreign exchange (another meaning of basic industries – those that generate exports).

Because Tanzania had its own version of socialism and was not aligned with the Soviet bloc, the World Bank was willing to support this strategy. One project that was supported with a World Bank line of credit was the Morogoro Shoe Factory in Tanzania, which never operated at more than 7 per cent of capacity and was meant to export shoes to Italy. Quite why Italians would need shoes from Tanzania was not clear, as the country is not noted for its lack of high-quality footwear, and the World Bank engaged in historical amnesia, with its staff critiquing projects such as this, even though the World Bank itself had funded it (Belli 1991). Thus, Tanzanian taxpayers were lumbered with debt for a poorly thought-through project. When I visited the plant in 1992 it was closed, but several people were employed to polish the machinery to ensure it wouldn't rust. The plant subsequently burnt down in 1997.

Like much of the Global South in the 1980s, Tanzania experienced a debt crisis, which was so bad at one point that it was reported that captains of oil tankers docking in the commercial capital, Dar es Salaam, refused to unload their cargo until suitcases full of hard currency, like US dollars, were delivered on board their ships – the Tanzanian government's credit had run out. Julius Nyerere famously referred to more capitalist Kenya as a "man eat man society". To which the Kenyan attorney general quipped that Tanzania was a "man eat nothing society". After protracted consultations and the establishment of an advisory group, Tanzania entered a harsh programme of economic austerity with the IMF beginning in 1986.

While the economy has grown relatively rapidly in recent years, it has been very natural resource intensive, particularly through exports of gold, for example. The country has also become highly import dependent, even for milk, though there is no reason Tanzania should not be able to produce sufficient quantities of this.

> An example for this practice is Bakhresa Group, one of the most successful business conglomerates in East Africa. Rather than taking risky investments into the dairy value chain in Tanzania, they import cheap EU milk powder to their plant in Zanzibar, mix it with stabilizers and water, and import it into mainland Tanzania where the product is sold as cheap "milk" (Ouma 2017, 11).

By some estimates the founder of this group, Said Salim Bakhresa, is a billionaire in terms of US dollars. Africa's "youngest billionaire", Mohammed Dewji, who is also from Tanzania, was kidnapped in 2018. Massive inequality has costs, even for the rich.

Case study: "socialism" and reform in Vietnam

The Vietnamese experience of "socialism" has been very different. Vietnam is technically a socialist country; however, when its economy faltered, in part because of what some economists called incentive problems, such as the soft budget constraint, it began a programme of economic liberalisation and reform called *doi moi* (renewal) in 1986 with the aim of creating a "socialist-oriented market economy". Inflation was over 400 per cent in Vietnam in 1986. High inflation affects those on fixed incomes, such as workers, more severely than those whose livelihoods depend on prices, such as business owners, who can raise them in response to, and also creating further, inflation.

Economic reforms always entail the creation of winners and losers, and *doi moi* in Vietnam resulted in economic restructuring and unemployment for some (Chossudovsky 1997). However, like in China, the economic reforms in Vietnam and its regional context resulted in rapid economic growth. As Henley (2015, 5) notes "in 1960, South-East Asians were on average much poorer than Africans; by 1980 they had caught up, and by 2010 there were two and a half time richer", with Malaysia by itself every year exporting more manufactured goods than all of sub-Saharan Africa by the end of the twentieth century.

Figure 2.3 displays trends in gross domestic product (GDP) for Tanzania and Vietnam. In the mid-1980s Vietnam had a substantially lower GDP per head than Tanzania; however, as its population was about three times as big, it had a bigger economy. In 1990 the economy of Vietnam was about 2.7 times the size of Tanzanian one, but by 2016 it was nearly four times bigger. By 2016 GDP per capita was more than double that of Tanzania.

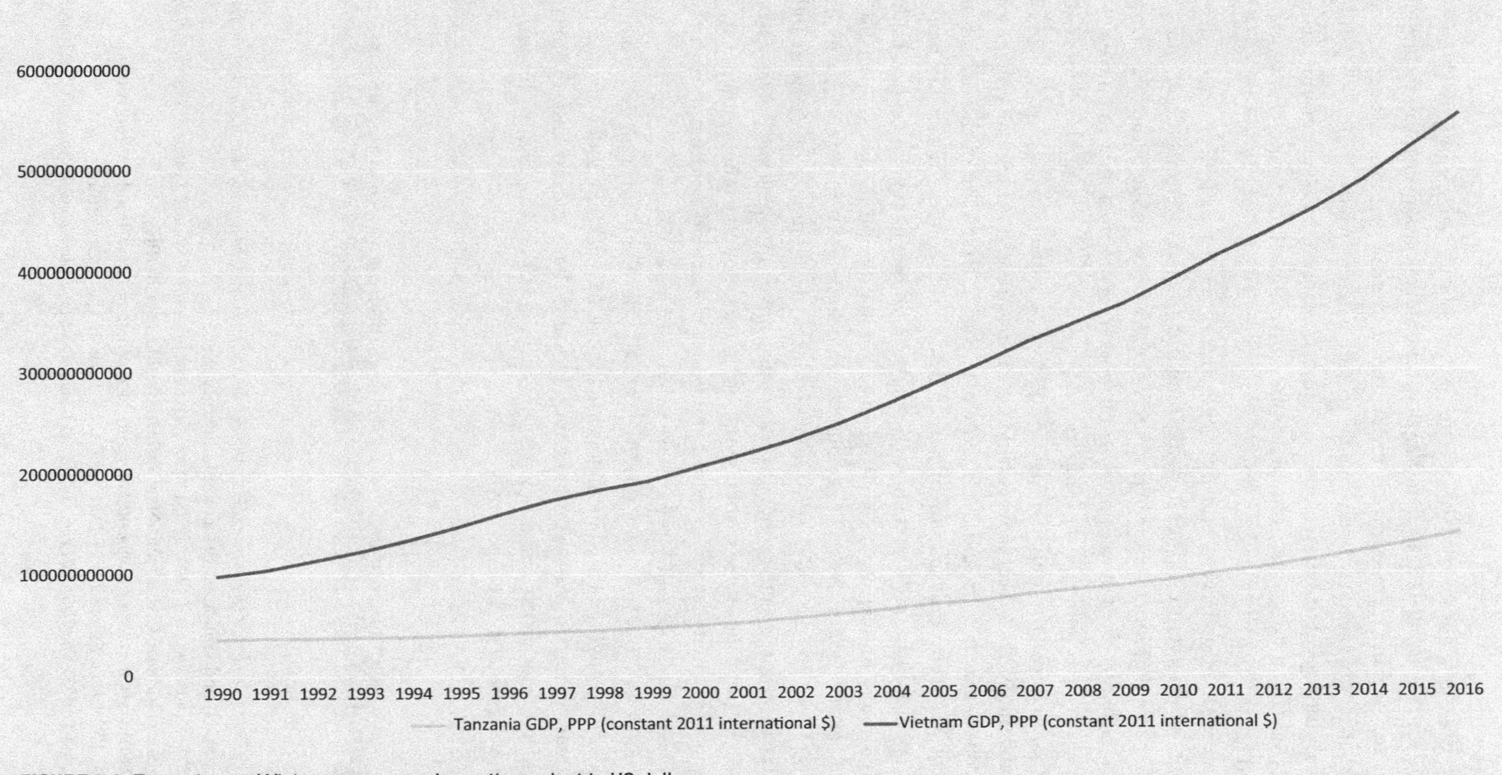

FIGURE 2.3 Tanzanian and Vietnamese gross domestic product in US dollars

Source: Calculated from World Bank (2017)

How did such a dramatic divergence come about? According to Henley (2015, 12) it was "state-led rural and agricultural development leading to higher incomes for peasant farmers, has been central to South-East Asia's success, and I infer that its absence has been crucial to sub-Saharan Africa's failure". While liberalisation of agricultural markets was important in Vietnam's economic take-off, it has also had an active industrial policy, including Five Year Plans (Altenburg 2011).

Despite widespread poverty and low wages in Africa, what are called "real product wages" are often lower in parts of Asia, meaning that foreign investors often prefer to locate there. Real product wage refers to the wage needed to produce one t-shirt, for example. Higher levels of development mean workers are more skilled and use better or more efficient technologies, keeping the real product wage down. Also, things like poor infrastructure across much of Africa or power cuts affect this too.

Despite the NAFC, Vietnam's garment exports grew at an annual average of 26 per cent between 2005 and 2010 (Newman *et al.* 2016). One of the ironies of the recent Vietnamese development is that despite the country notionally being socialist, as in China, authoritarian single-party regimes can repress or ban trade unions and thereby keep wages low to attract foreign investors. Thus, "socialist" states may actually offer the most conducive environment for private investors or capitalists. That said, the scale and pace of economic development in Vietnam and China has been rapid, although with substantial environmental and social costs. More than 80 per cent of Vietnamese think life in their country is better than it was 40 years ago. In these cases, "socialism" laid the groundwork and framework for rapid capitalist development – the reverse of what Marx envisaged.

Dependency theory inspired a variety of socialist development strategies around the world, most of which failed to achieve their goals or collapsed under the weight of their own dysfunctions and contradictions. However, has a shift towards the market provided the answer to the problem or problematic of development? This is the question we turn to in the next chapter.

Further reading

Books

Evans, P. 1979. *Dependent Development: The Alliance of Multinational, State, and Local Capital in Brazil.* Princeton, NJ and Guildford: Princeton University Press.

Gray, H. 2018. *Turbulence and Order in Economic Development: Institutions and Economic Transformation in Tanzania and Vietnam.* Oxford and New York: Oxford University Press.

Henley, D. 2015. *Asia-Africa Development Divergence: A Question of Intent.* London: Zed Books.

Rodney, W. 1973. *How Europe Under-Developed Africa.* Paris: Bogle-L'Ouverture.

Wilber, K. (ed.) 1973. *The Political Economy of Development and Underdevelopment.* Swedish, SL: Random House.

Articles

Cardoso, F.H. 1977. "The Consumption of Dependency Theory in the United States." *Latin American Research Review* 12 (3): 7–24.

Kiely, R. 2006. "US Hegemony and Globalisation: What Role for Theories of Imperialism?" *Cambridge Review of International Affairs* 19 (2): 205–21.

Website

Velasco, A. 2002. "Dependency Theory." *Foreign Policy* 133: 44–5, 48. https://search.proquest.com/docview/822409171?pq-origsite=gscholar.

References

Abrahamsen, R. 2000. *Disciplining Democracy: Development Discourse and Good Governance in Africa*. London: Zed Books.

Aglietta, M., and D. Fernbach. 1979. *A Theory of Capitalist Regulation: The US Experience*. London: NLB.

Altenburg, T. 2011. "Industrial Policy in Developing Countries: Overview and Lessons from Seven Country Cases." http://edoc.vifapol.de/opus/volltexte/2011/3341/pdf/DP_4.2011.pdf.

Amin, A. (ed.) 1994. *Post-Fordism: A Reader*. London: Wiley.

Amin, S. 2016. Keynote Address at *Political Economy: International Trends and National Differences*. International Initiative for the Promotion of Political Economy, School of Economics & Management, University of Lisbon, September.

———. 2006. "The Millennium Development Goals: A Critique from the South." *Monthly Review-an Independent Socialist Magazine* 57 (10): 1–15.

Amsden, A. 1989. *Asia's Next Giant: South Korea and Late Industrialization*. New York and Oxford: Oxford University Press.

Belli, P. 1991. "Globalizing the Rest of the World." *Harvard Business Review* 69 (4): 50–5.

Chen, J., J. Jing, Y. Man, and Z. Yang. 2013. "Public Housing in Mainland China: History, Ongoing Trends, and Future Perspectives." In *The Future of Public Housing: Ongoing Trends in the East and the West*, eds. J. Chen *et al.* Berlin Heidelberg: Springer-Verlag.

Chossudovsky, M. 1997. *The Globalization of Poverty*. London: Zed.

De Janvry, A. 1981. *The Agrarian Question and Reformism in Latin America*. Baltimore, MD and London: Johns Hopkins University Press.

Evans, P. 1979. *Dependent Development: The Alliance of Multinational, State, and Local Capital in Brazil*. Princeton, NJ and Guildford: Princeton University Press.

Frank, A. 1967. *Capitalism and Underdevelopment in Latin America. Historical Studies of Chile and Brazil*. New York and London: Monthly Review Press.

Fukuyama, F. 1992. *The End of History and the Last Man*. New York: Free Press.

Global Footprint Network. 2015. "Only Eight Countries Meet Two Key Conditions for Sustainable Development as United Nations Adopts Sustainable Development Goals." www.footprintnetwork.org/2015/09/23/eight-countries-meet-two-key-conditions-sustainable-development-united-nations-adopts-sustainable-development-goals/.

Guardian, The. 2016. "European Parliament Slams G7 Food Project in Africa." www.theguardian.com/global-development/2016/jun/08/european-parliament-slams-g7-food-project-in-africa.

Harvey, D. 2003. *The New Imperialism*. Oxford and New York: Oxford University Press.

Henley, D. 2015. *Asia-Africa Development Divergence: A Question of Intent*. London: Zed Books.

Jackson, R. 1992. "Juridicial Statehood in Africa." *Journal of International Affairs* 46 (1): 1–16.

Jessop, B. 2002. *The Future of the Capitalist State*. Cambridge: Polity.

Kaplinsky, R., and C. Cooper. 1989. *Technology and Development in the Third Industrial Revolution*. London: Cass.

Kemp, T. 1972. "The Marxist Theory of Imperialism." In *Studies in the Theory of Imperialism*, eds. R. Owen and B. Sutcliffe. London: Longman, pp. 15–34.

Kiely, R. 2010. *Rethinking Imperialism*. Basingstoke: Palgrave Macmillan.

Kornai, J. 1992. *The Socialist System: The Political Economy of Communism*. Oxford: Clarendon.

Lenin, V. 1917 [2010]. *Imperialism: The Highest Stage of Capitalism*. London: Penguin Books.

Lipton, M. 1977. *Why Poor People Stay Poor: A Study of Urban Bias in World Development*. London: Temple Smith.

Mamdani, M. 1996. *Citizen and Subject: Contemporary Africa and the Legacy of Late Colonialism*. Princeton, NJ and Chichester: Princeton University Press.

Mandel, M. 2004. *How America Gets Away with Murder: Illegal Wars, Collateral Damage and Crimes Against Humanity*. London: Pluto.

Marx, K. 2011. *Das Kapital: Volume 3*. Creative Independent Publishing Platform.

Newman, C., J. Page, J. Rand, A. Shimeles, M. Soderbom, and F. Tarp. 2016. *Made in Africa: Learning to Compete in Industry*. Washington, DC: Brookings.

Ouma, S. 2017. *Africapitalism: A Critical Commentary*. Mimeo.

Pakenham, T. 1992. *The Scramble for Africa*. London: Abacus.

Peet, R., and E.R. Hartwick. 2009. *Theories of Development: Contentions, Arguments, Alternatives*. 2nd edition. New York and London: Guilford.

Pew Research Centre. 2014. "Most Americans Think the U.S. Is Great, but Fewer Say it's the Greatest." www.pewresearch.org/fact-tank/2014/07/02/most-americans-think-the-u-s-is-great-but-fewer-say-its-the-greatest/.

Riddell, R. 1990. *Manufacturing Africa: Performance and Prospects of Seven Countries in Sub-Saharan Africa*. London: James Currey.

Schraeder, P. 2000. *African Politics and Society: A Mosaic in Transformation*. Basingstoke: Palgrave Macmillan.

Sender, J., and S. Smith. 1986. *The Development of Capitalism in Africa*. London: Methuen.

Shaw, M. 2003. "The State of Globalization: Towards a Theory of State Transformation." In *State/Space: A Reader*, eds. N. Brenner, M. Jones and G. MacLeod. Cambridge and Malden, MA: Wiley-Blackwell.

Siakwah, P. 2017. "Are Natural Resource Windfalls a Blessing or a Curse in Democratic Settings? Globalised Assemblages and the Problematic Impacts of Oil on Ghana's Development." *Resources Policy* 52: 122–33.

Smith, J. 2018. "David Harvey Denies Imperialism." http://roape.net/2018/01/10/david-harvey-denies-imperialism/.

Sullivan, M.P. 2001. *Theories of International Relations: Transition vs: Persistence*. Basingstoke: Palgrave Macmillan.

Thomas, C. 1974. *Dependence and Transformation: The Economics of the Transition to Socialism*. New York and London: Monthly Review Press.

Wallerstein, I. 1975. *The Capitalist World Economy*. Cambridge and New York: Cambridge University Press.

Warren, B. 1980. *Imperialism: Pioneer of Capitalism*. London: Verso.

Woodley, D. 2015. *Globalization and Capitalist Geopolitics: Sovereignty and State Power in a Multipolar World*. Oxford and New York: Routledge.

CHAPTER

3

The neoliberal revolution in development

Participation, power and poverty

> [After the Second World War] It does suggest that the earlier policy aim for Europe was reincorporation into the capitalist world economy. In contrast, the aim for the Global South in the 1980s and 1990s was more akin to transformative discipline. . . . It is in this sense that the restructuring programs were more about debt service: they aimed at shaping a political economy and a repositioning of these countries as sites of extraction, ranging from natural resources to the consumption power of their populations.
>
> – (Sassen 2014, 90)

The genesis of neoliberalism

In a capitalist society business must be profitable for there to be jobs, goods and services produced and taxes paid – that is, for the society to reproduce itself. Maintaining, enhancing or restoring business profitability then becomes a key role for the state in a market economy based on private property ownership. This means the state is not class neutral in capitalist societies, but largely works to protect and expand business interests, as this is a prerequisite for the reproduction of the society or social formation. However, depending on the balance of class power, the state of development of technology (which is partially a product of the nature of class power), public opinion and other factors, there may be different ways for business to achieve profitability. As noted earlier, wages are contradictory in capitalist economies as they are both a cost for business and a source of demand, as the majority of people, at least in rich countries, earn their living from them.

The post-war economic boom in the US and Europe, known as *les trente glorieuses* (the 30 glorious [years]) in France, had been underwritten by a class compromise, sometimes called "Fordism", after Henry Ford who founded the famous American car company. As noted earlier, Ford had realised that wages were not just a cost of production for business but could also serve as a source of demand for the products of industry. This insight was reproduced in policy across the developed world through

Keynesianism. Keynes (1936), the famous British economist, argued that if another Great Depression was to be prevented, and war which followed on the heels of it, then governments needed to engage in active demand management and not just let the "free" market determine the course of the economy, as if unregulated, it was subject to cycles of boom and bust.

The system which was put in place in 1944 internationally facilitated this type of economic approach – the Bretton Woods Agreement. The Bretton Woods Agreement pegged the value of different participating currencies around the world to the price of gold. The goal here was to prevent the kind of competitive "beggar thy neighbour" currency devaluations that were said to have worsened the Great Depression. Reducing the value of your currency makes it cheaper for foreigners to buy your products and more expensive to buy imports, thereby often improving trade balance, but at the cost of rising inflation as imports become more expensive. If we take the example of the Eurozone in Europe and the United States, we can see how this works. If the Eurozone were to devalue its currency the United States would suffer, as it would make American exports into the zone more expensive and exports from Eurozone members to the US cheaper, so consumers there would buy more of them. The Bretton Woods system was underwritten by the United States and its economy, which after the Second World War accounted, by some estimates, for fully half of global economic output, given the devastation in Europe and Japan following the war.

The Bretton Woods system allowed countries to keep extensive capital controls in place. However, as the German and Japanese economies boomed after the Second World War and their products became more competitive, the United States began to run substantial trade deficits with them, draining the US of gold (liabilities). This put pressure on American corporate margins, as did the fact that its companies tended to pay relatively high wages in global terms. During the "stagflation" of the late 1960s and early 1970s the American government had sought to stimulate the economy by printing money, thereby creating demand, but much of the demand for products was being met by overseas imports, and printing money often stokes inflation. Milton Friedman (1970, 24) argued that "inflation is always and everywhere a monetary phenomenon in the sense that it is and can be produced only by a more rapid increase in the quantity of money than in output". However, inflation can also be caused by input scarcity, making their prices more expensive – so-called "cost push" inflation.

The situation became untenable for the US, and in 1971 President Nixon ended the US dollar's convertibility to gold. This action effectively ended the Bretton Woods Agreement and set the stage for the current era of financialisation, where profits and flows of speculative investments can largely determine government policy, thereby structuring the broader functioning of the economy. Without capital controls, if governments raise taxes on capital, it will, all other things being equal, flow out of the country, giving a major incentive for governments to be "business-friendly" under neoliberalism. The fact that the Keynesian policy of trying to boost demand proved to be unsustainable led to a shift in what is sometimes called the regime of accumulation: from Fordism to neoliberalism, which is also sometimes known as a regime of flexible accumulation. A regime is a set of rules about how something works.

There were other important developments in the following years, which were to dramatically affect the US economy. In 1973 the Organization of Petroleum Exporting (OPEC) countries imposed an oil embargo on the United States in response to its support for Israel in the co-called Yom Kippur War, when a number of Arab states launched a surprise attack on that country on Judaism's holiest day. This led the price of oil to quadruple globally and fed inflation and recession in the US in 1974–1975. A second oil price shock was induced by the Iran-Iraq War of 1980–1988, which reduced those countries' oil output, again feeding inflation in the US. Perhaps somewhat ironically in light of later history when the US invaded Iraq to depose him, the United States supported the Iraqi dictator Saddam Hussein in this effort against the newly installed regime in Iran which held American diplomats and citizens hostage for more than a year in the American Embassy in Tehran. The response of the US Central Bank under Paul Volker to this increased inflation was to drastically raise interest rates to reduce the money supply in order to try to slow it and attract capital into the country. This became known as the "Volker shock".

As many developing countries had their loans denominated in US dollars, this often meant they had to pay higher interest rates on them. Many developing countries had gone heavily into debt during the 1970s as Middle Eastern countries in particular "recycled" their "petro-dollars" in the form of loans to them, as their own economies did not have the capacity to productively invest them. Also, as the value of the US currency went up or appreciated with higher interest rates in the US, this made repaying dollar-denominated loans in local currency terms more expensive.

The upshot of all of these developments for much of the Third World was a full-blown debt crisis, which erupted when Mexico announced it could no longer service its international debts in 1982 (Korner 1985). This sent shock waves through the international financial community and system and brought the credit-worthiness of many other developing countries into question. Consequently, international banks largely stopped lending to them, creating a balance-of-payments crisis for many countries, where they had extreme difficulty getting the foreign currency they needed to import essentials to keep their economies running, such as oil. The balance of payments refers to trade and investment flows and is composed of two elements: the current account and the capital account. Despite what is sometimes written, there can never be a deficit in the balance of payments, as the current and capital accounts must always balance each other out. For example, if a country imports more than it exports, it must import capital to pay for "excess" imports. However, there may be deficits or surpluses on either the current or the capital account.

The theory and roll-out of neoliberalism

The stagflation of the 1970s in much of the West was to have important implications around the world. The Conservative Margaret Thatcher was elected prime minister in the United Kingdom in 1979 and conservative Republican Ronald Reagan as president in the United States in 1980. Both saw it as their mission to revive the economies of their countries by undermining the power of organised labour, thereby reducing wages and restoring profitability for business. However, this would also depress general demand, and so the other leg of restructuring was required – opening markets around the world.

Neoliberalism is the term generally used by critics of "free market" policies to describe them, although a recent paper in an IMF journal was titled "Neoliberalism: Oversold?" (Ostry *et al.* 2016), thereby implicitly accepting some of the critique while still standing by its fundamentals. Some also argue that the term has been so overused that consequently it has been virtually evacuated of meaning. If everything is seen as neoliberal and if everything is neoliberal, the term loses its analytical purchase, so in the end nothing is neoliberal, it is argued.

The term liberal has different meanings in Europe and the US, where it connotes people who have left-of-centre political leanings. However, in the UK Margaret Thatcher, who was prime minister from 1979 to 1991, was considered an economic liberal. The root of the word liberal comes from liberty, meaning freedom – hence the "free" market. "Economics is the method", Margaret Thatcher declared, but "the object is to change the heart and soul" (quoted in Klein 2014, 60), so neoliberalism was and is fundamentally about changing the way people think about themselves and interact with others. The supposedly inherently selfish nature of people is meant to promote wider social welfare. Conventional economics posits that people act as rational self-interested profit and happiness (what economics calls utility) maximisers – *Homo economicus* (economic men). This is a foundational belief in neoliberal thought, and Treanor (n.d.) argues that neoliberalism is "an ethic in itself, capable of acting as a guide to all human action, and substituting for all previously held ethical beliefs". Being selfish is supposedly good for other people – or in the words of the fictional stock broker Gordon Gekko in the film *Wall Street* – "greed is good". However, perhaps his name is meant to be ironic and connote his inhuman qualities, as a gecko is a type of lizard.

Given the scale of global inequality described earlier, however, to what extent can markets be said to be "free"? If you don't have land or other assets and the economy is poor, to what extent are you free? You may be "free" (compelled by necessity) to sell your labour for often very low wages or go into debt and be exploited. If there are no jobs, you may be "free" to live in poverty or starve. In this way exploitation or exclusion can be presented as freedom – a strange form of what George Orwell called "doublethink" (Brooks 2017). Such thinking was mirrored in Margaret Thatcher's comment that there was "no such as thing as society", only "individual men and women" and families, so there are assumed to be no reciprocal obligations or solidarity amongst and between people, only competitive individualism. At its core neoliberalism is the belief that most social interactions and exchanges are best coordinated through the market (Harvey 2005). But what is new (the neo in neoliberal) about economic liberalism, which was after all largely fashioned in the United Kingdom during the eighteenth and nineteenth centuries?

Adam Smith, who is often cited as the founder of free-market economics, believed that the "invisible hand of the market" could optimise social welfare through the aggregation of individual decisions. It was bakers' or other producers' greed, not love of other people, which would drive them to produce bread, for example, to meet social need, he argued ("It is not from the benevolence of the butcher, the brewer, or the baker that we expect our dinner, but from their regard to their own interest"). However, this greed could be destructive if it led to concentrated market power, as producers could gouge customers if there weren't alternative sources of supply. So it would need to be kept in check by competition, he argued. If there were multiple bakers and few barriers to entry, there would be incentives to produce high-quality bread at an affordable price, for example. This suggested that the state should play an active role in competition policy, ensuring that monopolies[1] (where there is only one provider) and monopsonies (where there is only one buyer) did not emerge. What was neo about neoliberalism was that in the face of rising competition from the Japanese and German economies, Ronald Reagan's administration believed that it should encourage the emergence of US-based companies with global market power. This was successful in some sectors, particularly in high technology. In 2018 the Microsoft Windows operating system was used on 83 per cent of desktop computers worldwide (Statista 2018). It has been reported that in China, for example, Microsoft did not enforce its intellectual property rights until its products had become the operating standard in that country (Patel 2009).

There are also some other differences between classical liberals and neoliberals. Whereas eighteenth- and nineteenth-century liberals did not believe in (strong) central banks or intellectual property rights, neoliberals generally believe in these (Chang 2018). They often believe that central banks should be "independent" – that is, that they should have the power to set interest rates independent of political concerns or the opinions of the public. Some have argued that these types of policies have led to a situation where "corporations rule the world" (Korten 2001), as "independent" central banks pursue policies that are favourable to transnational capital in order to attract investment. Consequently, globalised neoliberalism is reflective of what some call the structural power of transnational capital, or its ability to set the rules of the international economic "game". In a sense, the term game downplays the importance of what is at stake, as the intense competition in the global economy represents a form of what some have called structural violence, where widespread poverty and exclusion result in countless people's deaths, misery and malnutrition around the world. However, it does capture the idea that there are winners and losers in the competition.

Under the current global free-market system, if governments pursue policies that attempt to restrict the movement of capital or to tax or regulate it in other ways, international investors will go on "strike" (i.e. refuse to invest in that country). As noted earlier, if governments adopt policies that financial investors dislike, such as raising taxes on capital, they will move their money out of the country, perhaps creating an economic panic or crisis or resulting in downgrades by ratings agencies, such as Moody's, of government debt, thereby raising the interest rates governments have to pay. In this way, the structural power of financial capital under neoliberalism is manifest.

The economic theory behind neoliberalism is based on so-called neoclassical economics. This terminology again requires some definition. "Classical" generally is taken to imply that something is old, so neoclassical economics is "new-old" economics. Neoclassical economists accept some of the central tenets of older schools of economics, such as the "free market" and "free trade". However, one of the things that makes it "new" is its mathematisation. In its attempt to mimic the hard sciences, neoclassical economics is sometimes critiqued for being inscrutable to outsiders because of the nature and extent of its mathematisation. This, however, serves an important political function – it means that critics who cannot engage its mathematisation are often dismissed by its practitioners as not understanding how economics works. It is somewhat ironic that the development of neoclassical economics and its mathematisation has reached its apogee in the United States but fundamentally misrepresents that country's history of economic development. Between its civil war in the nineteenth century and the Second World War, the US had the highest levels of economic protection in the world (Peet and Hartwick 2009), rather than engaging in free trade.

The roll-out of neoliberalism in the Global South

The elections of Reagan and Thatcher were to have momentous effects in the Third World, as they were to issue in the era of structural adjustment or economic liberalisation, sponsored and enforced primarily by the World Bank and IMF. Structural adjustment contained a variety of measures, however, among the main ones were increasing interest rates, reducing import tariffs and privatising SOEs. In Zimbabwe in the 1990s its economic structural adjustment programme (ESAP) was known colloquially as "extreme suffering of African people". The theory and practice of structural adjustment will be discussed in more detail a little later; however, it is worthwhile providing some more background on its emergence first.

Whereas the debt crisis in the Third World in the 1980s was sparked by the United States' central bank (called the Federal Reserve)[2] dramatically raising interest rates in the late 1970s and early 1980s, this was not how it was represented in the IFIs. According to the World Bank, officials at the IMF and a number of academics, such as Anne Kruger, who coined the term rent-seeking and worked at both the World Bank and the IMF at different times, the economic crisis in much of the developing world in the 1980s was an outcome of inappropriate economic policies, which protected inefficient domestic industries and were biased against rural development and exports. It was argued that these policies were designed and perpetuated, not because of their inherent public merit, but because of the rent-seeking (opportunities for corruption) they afforded state elites and their cronies. In order to discipline the state and force "economic efficiency", it was argued that privatisation, trade liberalisation, monetary tightening or raising interest rates and other policy reforms should be instituted. These policies, which became the prevailing "common sense" in policy circles and circuits across much of the world, were collectively known as the "Washington Consensus", as they were most ardently promoted by the US Department of the Treasury, World Bank and IMF. However, this neoliberalisation was paradoxical in that it was a process of the global deregulation of capital through the international regulation of states, but indirectly driven by the world's most powerful states at the time, such as the US, through its voting weight and influence in the IFIs.

As noted earlier, in order to make their models work, neoclassical economists make all sorts of unrealistic assumptions, such as there being no unemployment. In neoclassical economics, the market is seen to be the best system of socio-economic organisation. However, it is acknowledged that markets can sometimes fail and that consequently there may be a restricted role for the state in economic development. For example, firms may underinvest in their workers' skills or education because their workers may leave and take their skills with them. This is often used as a justification for state involvement in the education sector, as education is considered to be a public good, as noted earlier. If the state provides or subsidises education, then it will not be undersupplied to the same extent as it might be if it was left in the hands of the private sector.

Whereas many economic models are based on unrealistic assumptions, Milton Friedman (2008), one of the doyens of the neoclassical Chicago School of Economics, famously said that the realism of assumptions didn't matter and what was important was the ability to predict outcomes. What goes unasked and unanswered in such a proposition is whose interests are served by certain assumptions. For example, the theory of free trade generally serves economically stronger countries' interests. The former US trade representative under George Bush Sr., Carla Hills, once advocated that the US should pry open countries with a crowbar "so that our private sector can take advantage of them" (quoted in Dunkley 2004, 220). The "crowbar" included the structural adjustment programmes (SAPs) of the World Bank and IMF implemented throughout much of the Global South during the 1980s and 1990s, with often devastating economic and social consequences. For example, exposing infant industries to full-blown competition often resulted in their destruction across much of Africa, Latin America and parts of Asia.

For their proponents "free" markets enhance social welfare by ensuring efficient allocation of resources. Even in the presence of market failures, some of the staunchest neoliberals argue that government failure can be worse, and consequently the less government "interference" in the market, the better. In neoliberal monoeconomics, where it is assumed the same economic "laws" apply in all contexts, markets are assumed to be perfect – that is, that there are perfect information flows about prices and quality of products, and consequently that they represent the best and most efficient system of social organisation possible. It is a common strategy amongst neoliberal economists to argue that where "development" is failing it is because "functioning market systems in the 'Global South' are deemed 'imperfect' by comparison to text-book 'perfect market' processes of resource allocation, distribution and utilization" (Selwyn 2014, 21).

Prior to the "neoliberal revolution", the international development regime had been relatively permissive of different policy approaches (Amsden 2007), partly for geopolitical reasons – to make the Soviet alternative less attractive. However, with "stagflation", or a combination of stagnation and rising prices (inflation), and the oil shocks of the 1970s in the advanced economies, the imperative in the major industrial states became to raise the rate of profit through wage cost reduction and the opening of new export markets through SAPs in the Third World (Bello *et al.* 1994), discussed in more detail later.

While neoliberalism as a form of praxis is often associated with Thatcher and Reagan, it was initially put into practice in Chile under the dictatorship of General Augusto Pinochet (Harvey 2005). General Pinochet came to power in a violent army coup in 1973 in which the democratically elected socialist government of Salvador Allende was overthrown. During the coup, on 11 September, the Presidential Palace was bombed by the air force. The change of power was supported by the US government. Previously

President Nixon in the US had instructed the Central Intelligence Agency (CIA) to make the Chilean economy "scream".

Milton Friedman, one of the main founders and promoters of neoliberalism, was a supporter of General Pinochet and his brutal rule in which thousands of people were "disappeared". The so-called Hayek-Friedman hypothesis contends that politically free societies must have high levels of what they determined was economic freedom (Lawson and Clark 2010). In practice, Friedman was happy to support Pinochet, as was Margaret Thatcher, and talked about the "miracle of Chile". According to Hayek (1978 quoted in Selwyn 2017, 84), "I have not been able to find a single person even in the much-maligned Chile who did not agree that personal freedom was much greater under Pinochet than it had been under Allende". Perhaps they were afraid to say if it wasn't or had been murdered.

Under Pinochet economic policy making was dominated by the so-called "Chicago Boys", a group of economists who had trained with Friedman and others at the University of Chicago. They pursued radical policies of economic liberalisation, with some notable exceptions, such as the maintenance of the copper company, Codelco, in state hands, which provided valuable resources and financing for the regime. Once neoliberal policies were "road-tested" in Chile, they were then later rolled out by the Reagan and Thatcher governments in their home countries and then to much of the world through World Bank/IMF SAPs. Thatcher was also influenced by anti-tax movements in the United States, and the US played a particularly important role in the role-out of neoliberalism in the Global South.

The international financial institutions

As noted earlier, the United States has a special place in the governance of the world's two most important international financial institutions: the World Bank and the IMF. According to the World Bank's website:

> The United States was a leading force in the establishment of the World Bank in 1944 and remains the largest shareholder of the World Bank today. As the only World Bank shareholder that retains veto power over changes in the Bank's structure, the United States plays a unique role in influencing and shaping development priorities.
>
> (World Bank 2017e)

The World Bank and IMF are specialised agencies of the United Nations but operate differently from its other arms. Whereas in the United Nations' General Assembly the principle is "once country, one vote", in the IFIs it is "one dollar, one vote" based on contributions or subscriptions. While there have been questions raised as to whether China's vote, with that country having nearly a fifth of the world's population, should be weighted the same in the UN General Assembly (GA) as that of a small-island developing state, with a population of thousands, is partly academic. The votes of the GA may carry moral force but lack an enforcement mechanism, although the US Department of State can cut aid to countries that vote against it on issues it designates as important in the GA. So much for international democracy. The same cannot be said of the UN Security Council, which can authorise military action, for example – where the "great powers" of the Permanent Five (P5 – the United States, China, the UK, France and Russia) have vetoes, or of the World Bank and IMF, which have financial force, both directly through their loans and by signalling to private capital whether or not states are adopting "investor-friendly" policies. If you don't do what they want, you won't get access to loans and capital from them.

The American magnates, the Rockefeller brothers, gifted the land in Manhattan to the UN to ensure that it would be headquartered in the US in perpetuity (Hanahoe 2003), but a democratic deficit is even more starkly in evidence in the governance of the IFIs. Global power differentials find expression in the World Bank and IMF in that they reflect current and historical distributions of economic and, consequently, political power, and this is reflected in their history and geography. They were both founded at the Bretton Woods Conference in New Hampshire in 1944 and as a result are sometimes known as the Bretton Woods Institutions. They are both headquartered in Washington, D.C., close to the White House and the US Department of the Treasury, which substantially controls their purse strings.

Even if we were to accept that economic weight should be related to voting rights – that is that the rich should have more voting weight than the poor (which is an extremely problematic position from a moral or normative perspective) – there are substantial discrepancies in current weightings. For example, Belgium has a voting weight of 1.61 per cent in the World Bank, while Brazil's voting weight is 1.83 per cent (World Bank 2016). Belgium has a population of 11 million people, whereas Brazil has a population of 211 million and an economy four times bigger than Belgium's (World Bank 2017b).

The official title of the World Bank is the International Bank for Reconstruction and Development. It was initially set up to help rebuild Europe after the devastation of the Second World War. As the United States has disproportionate influence on it, it tends to follow and mirror politico-economic developments in that country closely. The president of the World Bank has always been an American citizen, while the managing director of the IMF has always been a European, although there was controversy during recent selection processes, as it was suggested that these appointments should be merit based. However, this proved to be too radical a departure, and geopolitical power has, to date, trumped justice or meritocracy in these selection processes.

When modernisation theory was in vogue in the post-war period the World Bank followed its prescriptions. Perhaps the best-known, or iconic, president of the World Bank was Robert McNamara, who served in that capacity from 1968 to 1981 before the neoliberal (counter)revolution at that institution. He had previously served as Secretary of Defense and at the Ford Motor Company. McNamara believed in planning at Ford and at the Department of Defense, where he oversaw the American war effort in Vietnam, and at "the Bank", as it is known colloquially. During his tenure "the World Bank became more concerned with income distribution, basic need and poverty reduction. Lending to infrastructure fell to only 36%, while the allocation to agriculture and social areas rose to 41% compared to 17% in the 1960s" (Stein *et al.* 2018, 4). However, this was to change after his departure. The publication of a report on "Accelerated Development in Sub-Saharan Africa: An Agenda for Action" (World Bank 1981) signalled the new priority to be given to SAPs in World Bank policy.

State power and neoliberalism

Given that the United States is the world's largest economy, it is the world's most powerful state in terms of the structural power it wields in the construction of international institutions, like the UN and regimes, or international sets of rules. However, there are also challenges and changes to this, given the contradictions or dialectics (interpenetration of opposites) of globalisation. Whereas the global sale of commodities is commonly thought to be separate from the state, governments raise taxes on labour and capital during the production, transfer and exchange moments in the circuit of capital, where money is converted into capital and, as long as the business is profitable, more money, which is then reinvested – hence a circuit. This means that commodity production and widespread circulation is also constitutive of state power. Thus, power of state and capital or business are partly inter-constitutive, with the context determining which fraction of capital is most influential or in the ascendancy. With the hollowing out of the US economy, partly as a result of competition from China, it has become increasingly dependent on its financial sector. However, this is not without its own contradictions, as evidenced by the NAFC. Nonetheless, and somewhat paradoxically as discussed earlier, neoliberalism remains dominant in international policy circles.

Neoliberalism is often thought to be the antithesis of industrial policy, as it "kicks away" the developmental ladder after the currently rich or developed countries have climbed it (Chang 2002). However, an alternative way to conceptualise the phenomenon is that neoliberalism is an industrial policy for the West (Tandon 2015). In particular, it is designed to reduce labour costs for transnational capital and open up resource and market access for Western corporations. Neoliberalisation, while not reducible to it, is partly a form of "central planning from Washington" (Abugre 2003), or a strategic industrial policy which is in part instantiated through the diffusion and growth of GVCs and production networks, which are planned and securitised through logistics and other mechanisms (Cowen 2014). We can see the recurrence of historical patterns here. As Friedrich List (1856 quoted in Selwyn 2014, 29), arguably the father of industrial policy, noted at the time when it was the "workshop of the world":

> a country like England, which is far in advance of all its competitors, cannot better maintain and extend its manufacturing and commercial industry than by a trade as free as possible from all restraints. For such a country, the cosmopolitan and the national principle are one and the same thing.

These types of policies have implications for sovereignty or ultimate authority in particular places. For example, as Sassen (2014, 85) notes in relation to the impact of SAPs on land access, "these diverse programs had the effect of re-conditioning national sovereign frameworks in ways that enabled the insertion of national territory into the new or emerging global corporate circuits. Once there, territory became land for sale on the global market". However, some developing countries have retained autonomy from the IFIs. For example, the Chinese government employs thousands of people to monitor Internet traffic in that country and also bans the use of certain websites such as Facebook, along with thousands of others. The "Great Chinese Firewall" is partly about censorship; it is also a form of digital protectionism which has allowed Chinese social media and other companies to grow by preventing access to that market by American-based companies, such as Facebook (Mann 2016).

In the wake of the debt crisis, the World Bank and the IMF were willing to provide balance-of-payments support for developing countries if required. However, these loans came with conditions. The IFIs claimed these conditions were necessary to ensure economic recovery, thereby enabling their loans to be repaid. The conditions became codified in SAPs which required governments to dramatically cut back on their expenditures, privatise industries, and liberalise trade so that so-called comparative advantage could operate.

Cutting back government expenditure was meant to serve a variety of purposes. It was designed to reduce demand in the economy and thereby reduce inflation. Also, by compressing demand, the demand for imports was meant to be reduced, thereby contributing to balance-of-payments stabilisation. Furthermore, in neoclassical economic theory, markets are meant to equilibrate, or clear, if there are no "distortions" in the economy. In this way of thinking, if there is unemployment it is because of government regulation, and interference, such as minimum wages, makes workers too expensive to employ, the argument goes. The solution proposed then is to make wages lower to make workers more attractive for employers to give jobs to. Laying off government workers was meant to increase unemployment and drive down the price of labour, thereby supposedly stimulating the economy after initially inducing a recession.

One of the problems with this approach is that people are not commodities. While it may be possible to sell more coffee or steel by reducing their prices, if the price of the labour is driven below subsistence, people can't work as they are malnourished, unhealthy, etc. Consequently, labour is what Karl Polanyi (1944) called a "fictitious commodity", which is not produced through market mechanisms, but in the case of labour through the family. Treating people as if they are a commodity leads to all manner of human suffering and social dysfunctions, including higher crime, for example, as people are pushed into poverty.[3] In Africa many civil servants were forced into multiple modes of livelihood, such as running businesses on the side or urban farming, which took them away from their primary occupations. A recent study found that 80 per cent of police officers' income in the capital of the Democratic Republic of the Congo comes from extorting bribes from motorists (Economist 2018).

The term commodity can be used in two ways. In a Marxian sense a commodity is something produced through and sold on the market. The second way in which the term commodity is used is as something being unprocessed or semi-processed such as iron, steel or cocoa. Structural adjustment also interacts with other economic tendencies, such as the long-term tendency for the price of raw materials and primary commodities to fall. For example, in 2007 prices for cocoa in real (inflation-adjusted) terms were less than a third of what they had been 30 years previously (Food and Agriculture Organization of the United Nations 2009). This is a major problem for big cocoa producers such as Ghana and Côte D'Ivoire in West Africa. While cocoa farmers were producing substantially more, under the incentives created by World Bank/IMF SAPs to export, as noted earlier, the income elasticity of demand for primary commodities is low. When supply increases, prices fall because people in rich countries do not buy, for the most part, two bars of chocolate or two bananas instead of one when their incomes go up, by way of example. On the other hand, people living in poverty spend much of their income on food; sometimes up to 50 per cent of their total (Spence 2011).

In relation to cocoa the effect of free-market programmes was to make West African farmers "run" faster to go backwards: that is, they were producing more, but prices were falling faster, meaning they got less return for putting in more effort. More labour power and environmental resources are being expended to achieve reduced returns for farmers, while consumers, mostly in rich countries and TNCs, get cheaper products and/or higher profits. This can be exacerbated by other tendencies. For example, in 2017 cocoa prices fell dramatically on the heels of a bumper harvest and reduced demand for chocolate as consumers' health consciousness in the developed world increased, according to media reports.

Many commodities around the world, such as wheat, oil or copper, are traded in US dollars. This means many countries have to maintain government reserves primarily in dollars, meaning the US can print its money and buy things from overseas effectively for free – something known as the privilege of seigniorage. When Britain played this role in the global economy, the famous British geographer Halford MacKinder said this enabled the UK to "share in the activity of the brain and muscles of other countries".

The long-term tendency of the price of primary commodities to fall and for the price of manufactured goods to rise results in a "trade trap" for raw materials exporters (Coote and LeQuesne 1996). In order to finance their balance-of-payments difficulties countries often resort to debt from abroad to pay for "excess" imports that they can't pay for with the dollars they have earned from exports. When this pattern is continued through time, it results in a "debt trap" (Payer 1975), as countries may not be able to repay their loans. These debts result in a further flow of profit back to generally developed countries where many of the biggest multi-national banks are headquartered. There are also many other mechanisms through which surplus is extracted from the periphery of the global economy to the core. For example, "when entrepreneurs in developing countries borrow money from abroad, for example, the requirement that their own state should have sufficient foreign exchange reserves to cover their borrowing translates into the state having to invest in, say US Treasury bonds. The difference between the interest rate on the money borrowed (for example, 12 per cent) and the money deposited as collateral in US Treasuries in Washington (for example, 4 per cent) yields a strong net financial flow to the imperial centre at the expense of the developing country" (Harvey 2005, 74).

If and when overextended developing countries can't pay, the World Bank and IMF step in as so-called lenders of last resort, but with conditions attached to their loans. The debt crisis in the 1980s, then, was a result of these structural conditions and the shorter-term or conjunctural event of the Volker shock.

The IMF and World Bank insist that debts to them and private financial institutions must be repaid unless there are agreed debt relief measures. In order to generate or save foreign currency to repay these debts, the IFIs insist on 1) export promotion and 2) import reduction ("compression"). Both of these policy prescriptions are deeply problematic, however. The first is problematic for a variety of reasons, including what economists call "the fallacy of composition". If the tendency is for primary commodity prices to fall, then encouraging countries to export more of these will only further depress prices – reinforcing the trade and debt traps. While it might be rational for say, Ghana to export more cocoa, as long as other countries didn't, when others such as Côte d'Ivoire and Brazil also export more, this will drive prices down considerably, perhaps producing a negative result for Ghana, even if more cocoa is produced. This is what the fallacy of composition refers to – that it is necessary to look at the global and not just the national picture when deciding economic policy. Selwyn (2014) also notes that this fallacy of composition can also apply to more statist theories which argue for strong government intervention in order to foster late industrialisation. What is good for one country – industrialisation – may not be good for others if it puts them under intense competitive pressure, which may result in deindustrialisation, for example.

A second aspect of economic "stabilisation" policy, import compression, is also problematic. One of the ways to reduce the amount of imports coming into an economy is to reduce the rate of economic growth. If people and businesses have less money, they will import less. This seems like a strange policy to promote for organisations like the IFIs whose remit is partly meant to be about promoting economic growth, and yet this is what they prescribe when they consider that "stabilisation" of an economy is required.

One way to reduce the level of demand in the economy is by raising interest rates (making the cost of borrowing money more expensive). This means that businesses invest less and import fewer machine tools, for example, and consumers buy less as the interest rates on their home and other loans and credit cards go up. Raising interest rates also reduces the level of inflation, all other things being equal, as there is less demand in the economy. Also laying off public-sector workers, partly to reduce government budget deficits, also reduces demand – another "benefit". Thus, World Bank/IMF programmes create austerity and may induce or worsen economic recession. However, this is only meant to be a short-term impact, but reduced investment has long-term effects, and increased primary commodity exports may not benefit the economy in the longer term, for the reasons outlined earlier. As Keynes noted "in the long term we are all dead".

Developed countries also have other policies to capture the value added from raw materials for themselves. One of these is so-called tariff escalation, noted earlier, where higher tariffs apply to manufactured or processed imports than to raw materials. According to the late Calestous Juma (2015), who was a professor at Harvard:

> Take the example of coffee. In 2014 Africa – the home of coffee – earned nearly \$2.4 billion from the crop. Germany, a leading processor, earned about \$3.8 billion from coffee re-exports. The concern is not that Germany benefits from processing coffee. It is that Africa is punished by EU tariff barriers for doing so. Non-decaffeinated green coffee is exempt from the charges. However, a 7.5 per cent

> charge is imposed on roasted coffee. As a result, the bulk of Africa's export to the EU is unroasted green coffee. The charge on cocoa is even more debilitating. It is reported that the "EU charges (a tariff) of 30 per cent for processed cocoa products like chocolate bars or cocoa powder, and 60 per cent for some other refined products containing cocoa".

Of course, the imposition of tariffs is in breach of neoliberal nostrums around free trade; however, the developed countries have been very successful in enforcing free trade in and on other weaker developing countries, while practicing protectionism when it suits their own interests. Neoliberal economists often rightly decry this double standard but are trumped by the political economy of power in this instance. However, where their prescriptions around the promotion of market opening in the Global South fit with prevailing distributions of power, they are often adopted.

Partly as a result of the economic channels noted earlier, combined with cutbacks in education and health care, structural adjustment was to have devastating economic and social consequences around the world. The raising of domestic interest rates in line with monetarist thinking and also to supposedly encourage businesses to adopt more labour, as opposed to capital-intensive production processes, reduced investment, often feeding into recession. Other austerity measures also contributed to this, and it was often the poor who were most affected as they could not afford to pay for private health care or education, for example. Fledgling or infant industries were exposed to global competition from more efficient producers in rich countries or NICs and often could not compete. When they closed, this also increased unemployment and reduced taxes (cf. Carmody 1998), furthering, rather than reversing, economic decline. The elimination of food subsidises often led to food or so-called IMF riots in many developing countries. Dramatic increases in the price of food in 2008–2009 and again in 2011 also led to riots in several countries around the world. Wheat prices in Egypt in 2008 were more than 150 per cent higher than their average from 1991 to 2007, feeding into discontent with the government and the subsequent "Arab Spring", when the government of long-term dictator Hosni Mubarak was overthrown. The Arab Spring resulted in the overthrow of a number of authoritarian governments in the region or plunged countries like Libya and Syria into civil war, with extensive external military intervention.

The policies of structural adjustment resulted in the so-called lost decade of the 1980s in Latin America and Africa, when many economic and social indicators for those regions deteriorated dramatically. This economic malaise extended to much of Africa into the mid-1990s, leading some to question whether it was becoming a "lost continent", although it is important not to overgeneralise and remember that Botswana in Southern Africa was the world's fastest growing economy from the 1960s through the 1990s (Samatar 1999). The *Economist* magazine in 2000 famously had a controversial and much-criticised cover about Africa entitled "The Hopeless Continent". What went unspoken in this was the role which neoliberalism played in the continent's decline. We now turn to a more in-depth discussion of the impacts of neoliberalism and whether or not there has been a revival of modernisation theory in more recent decades.

Notes

1 And also oligopolies, where there are only a few producers who collude to set or rig prices, did not emerge. However, the recent LIBOR (London Interbank Offered Rate) scandal revealed the extent of market concentration and power in financial services, as major banks colluded to falsify interest rates for their own benefit.

2 Interestingly, this is a partially private institution, as commercial banks hold stock in and can elect some of the board of governors of "The Fed".

3 In Western countries governments in fact try to keep a certain level of unemployment in the economy to prevent wage increases which might stoke inflation. The rate of unemployment where this is thought not to accelerate inflation is known as the NAIRU (non-accelerating inflation rate of unemployment). This represents a formalisation of what Marx talked about as the need to maintain a reserve army of labour for capital.

Further reading

Books

Bhagwati, J. 2007. *In Defense of Globalisation*. Oxford and New York: Oxford University Press.

Toye, J. 1993. *Dilemmas of Development: Reflections on the Counter-Revolution in Development Economics*. London: Wiley-Blackwell.

Articles

Bhagwati, J. 1982. "Directly Unproductive, Profit-Seeking (DUP) Activities." *Journal of Political Economy* 90 (5): 988–1002.

Das, R. 2015. "Critical Observations on Neo-Liberalism and India's New Economic Policy." *Journal of Contemporary Asia* 45 (4): 715–26.

Websites

The Mont Perlin Society, www.montpelerin.org

The World Economic Forum, www.weforum.org

The World Trade Organization, www.wto.org

References

Abugre, C. 2003. "Still Sapping the Poor: A Critique of IMF Poverty Reduction Strategies." http://chora.virtualave.net/sapping-thepoor.htm.

Amsden, A. 1989. *Asia's Next Giant: South Korea and Late Industrialization*. New York and Oxford: Oxford University Press.

Bello, W., S. Cunningham, and B. Rau. 1994. *Dark Victory: The United States, Structural Adjustment, and Global Poverty*. Pluto Press with Food First and Transnational Institute.

Brooks, A. 2017. *The End of Development: A Global History of Poverty and Prosperity*. London: Zed Books.

Carmody, P. 1998. "Neoclassical Practice and the Collapse of Industry in Zimbabwe: The Cases of Textiles, Clothing, and Footwear." *Economic Geography* 74 (4): 319–43.

Chang, H.-J. 2002. *Kicking Away the Ladder? Development Strategy in Historical Perspective*. London: Anthem.

———. 2018. *Address to the College Historical Society*, March 5th. Dublin: Trinity College.

Coote, B., and C. LeQuesne. 1996. *The Trade Trap: Poverty and the Global Commodity Markets*. New edition with additional material by Caroline LeQuesne. edition. Oxford: Oxfam.

Cowen, D. 2014. *The Deadly Life of Logistics: Mapping the Violence of Global Trade*. Minneapolis, MN: University of Minnesota Press.

Dunkley, G. 2004. *Free Trade: Myth, Reality and Alternatives*. London: Zed Books.

Economist, The. 2018. "Kinshasa's Traffic Police Make 80% of Their Income Informally." September 8th. www.economist.com/middle-east-and-africa/2018/09/08/kinshasas-traffic-police-make-80-of-their-income-informally.

Food and Agriculture Organization of the United Nations. 2009. *The State of Agricultural Commodity Markets 2009, High Food Prices and the Food Crisis*. Rome: Food and Agriculture Organization of the United Nations.

Friedman, M. 1970. "Counter-Revolution in Monetary Theory." Wincott Memorial Lecture, Institute of Economic Affairs, Occasional paper 33.

———. 2008. "The Methodology of Positive Economics." In *Philosophy of Economics: An Anthology*, ed. D. Hausman. Cambridge and New York: Cambridge University Press, 3rd edition, pp. 145–78.

Hanahoe, T. 2003. *America Rules: US Foreign Policy, Globalization and Corporate USA*. Dingle, CO: Kerry Brandon.

Harvey, D. 2005. *A Brief History of Neoliberalism*. Oxford: Oxford University Press.

Hayek, F. 1978. "Letter to." *Times of London*.

Juma, C. 2015. *How the EU Starves Africa into Submission*. https://capx.co/how-the-eu-starves-africa-into-submission/.

Keynes, J.M. 1936. *The General Theory of Employment Interest and Money*. London: Palgrave Macmillan.

Klein, N. 2014. *This Changes Everything: Capitalism Vs: The Climate*. London: Allen Lane.

Korner, P. 1985. *The IMF and the Debt Crisis: A Guide to the Third World's Dilemma*. London: Zed Books.

Korten, D. 2001. *When Corporations Rule the World*. 2nd edition. San Francisco, CA [Great Britain]: Berrett-Koehler.

Lawson, R.A., and J.R. Clark. 2010. "Examining the Hayek-Friedman Hypothesis on Economic and Political Freedom." *Journal of Economic Behavior & Organization* 74 (3): 230–9.

Mann, L. 2016. Presentation at *Connectivity at the Bottom of the Pyramid: ICT4D and Informal Economic Inclusion in Africa*, Rockefellar Centre, Bellagio, Italy, September.

Ostry, J., P. Loungani, and D. Furceri. 2016. "Neoliberalism: Oversold?" *Finance and Development* 53 (2). www.imf.org/external/pubs/ft/fandd/2016/06/ostry.htm.

Patel, R. 2009. *The Value of Nothing: How to Reshape Market Society and Redefine Democracy*. New York: Picador.

Payer, C. 1975. *Debt Trap: International Monetary Fund and the Third World*. New York: Monthly Review Press.

Peet, R., and E.R. Hartwick. 2009. *Theories of Development: Contentions, Arguments, Alternatives*. 2nd edition. New York and London: Guilford.

Polanyi, K. 1944. *The Great Transformation: The Political and Economic Origins of Our Time*. Boston: Beacon Books.

Samatar, A.I. 1999. *An African Miracle: State and Class Leadership and Colonial Legacy in Botswana Development*. Portsmouth, NH: Heinemann.

Sassen, S. 2014. *Expulsions: Brutality and Complexity in the Global Economy*. Cambridge, MA: Belknap and Harvard University Press.

Selwyn, B. 2014. *The Global Development Crisis*. Cambridge: Polity.

———. 2017. *The Struggle for Development*. Cambridge and Malden, MA: Polity.

Spence, M. 2011. *The Next Convergence: The Future of Economic Growth in a Multi-Speed World*. Crawley, WA: UWA Publishing.

Statista. 2018. *Global Operating Systems Market Share for Desktop PCs, from January 2013 to July 2018*. www.statista.com/statistics/218089/global-market-share-of-windows-7/.

Stein, H., S. Cunningham, and P. Carmody. 2018. *The Rise and Risks of the Random: The World Bank, Experimentation, and the African Development Agenda.* Mimeo.
Tandon, Y. 2015. *Trade Is War: The West's War Against the World.* New York and London: OR Books.
World Bank. 1981. *Accelerated Development in Sub-Saharan Africa: An Agenda for Action.* Washington, DC: World Bank.
———. 2016. *International Bank for Reconstruction and Development Subscriptions and Voting Power of Member Countries.* https://finances.worldbank.org/Shareholder-Equity/IBRD-Subscriptions-and-Voting-Power-of-Member-Coun/rcx4-r7xj.
———. 2017a. *Gross Domestic Product.* http://databank.worldbank.org/data/download/GDP.pdf.
———. 2017b. *The World Bank in United States.* http://www.worldbank.org/en/country/unitedstates.

CHAPTER

4

Impacts of neoliberalism and the revival of modernisation?

We have discussed the circumstances behind the adoption of neoliberalism as a set of policies, both in the developed and "developing world" contexts. In a sense it is not very helpful to talk about "the developing world", as there is only one world and as the discussion previously has shown, events in the core can dramatically affect the periphery or semi-periphery and vice versa, although often to a lesser extent, with some exceptions such as oil crises.

Another mechanism through which neoliberalism is transmitted to the Third World is the process of financialisation. Financialisation is driven both by the tendency towards overaccumulation in deregulated capitalist economies and the vast profits which can be derived from it. Overaccumulation as a concept can be explained by way of example. For example, workers in China's Foxconn plants producing iPhones and iPads for the global market cannot afford to buy the products they produce themselves, thus limiting demand. Apple's latest iPhone costs more than US$1,000. As a result, instead of profits being primarily earmarked for the construction of new factories, they often surge into other investments – such as stocks, property, and commodities – precipitating speculative bubbles and socially devastating busts, as seen in the case of the NAFC.

Financialisation may also become an embedded part of social formations, as in the United Kingdom, for example, where hundreds of thousands of jobs are dependent on the financial sector in the City of London. Administratively, the City of London is separate from the broader city, as it elects its own mayor and the electors are corporations (Urry 2014). As such, it represents a particular type of space of exception, which is administered differently from the rest of the territory. Consequently, there are spaces of convenience and exception which serve the needs of transnational capital, such as off-shore tax havens, in addition to spaces of exception which serve to securitise the current global formation, such as Guantanamo Bay in Cuba, where "enemy combatants" have been held for extended periods without trial (Mountz 2011).

Many states in the Global South are what Harrison (2004) calls "governance states" which implement the policies promulgated by the IFIs faithfully. At times local political elites may choose not to implement economic reforms demanded by the IFIs if alternative sources of funding are available from China, for example, or if the reforms are seen to compromise regime maintenance, or alternative courses of action open up greater opportunities for accumulation of assets; either political, monetary, physical or a combination of these. For example, in the 2000s Zimbabwe defied the IMF, but Robert Mugabe was able to promote legitimacy amongst his crony networks through redistribution of land as a strategy to stay in power (Carmody 2015). While Zimbabwe's "fast track land reform", or land invasions of Euro-Zimbabwean farms, and rejection of structural adjustment attracted the opprobrium of Western leaders, such as the UK prime minister at the time, Tony Blair, it was not judged to be a sufficient threat to Western interests to warrant military intervention. However, where there are strategic resources such as oil involved, this may not be the case, as the invasion of Iraq in 2003 and the NATO intervention in Libya in 2011 which led to the ousting of long-term president, Mommar Gaddafi, have shown (Carmody 2016). There were a variety of possible motivations for the invasion of Iraq; however, several US neoconservatives, who were subsequently to serve in the George W. Bush administration, had called for Saddam Hussein to be deposed in the Statement of Principles of the "Project for a New American Century" which is available on the World Wide Web. These included his vice president, Dick Cheney, and Secretary of Defense, Donald Rumsfeld.

In addition to neoliberalism there are also other axes through which major industrial powers seek to project their influence into the developing world, such as through military and "humanitarian" interventions, military bases and assistance, aid and diplomacy. Consequently, rather than speaking about US imperialism, for example, it is perhaps more useful to talk about what Kautsky (1914) called ultra-imperialism, where major powers coalesce to share sovereignty and power through institutions such as NATO, the EU and the World Bank and IMF, with the United States still being the "first amongst unequals". This creates a pattern where horizontal sovereignty sharing amongst the powers of the Global North is then projected southwards to create what has been termed elsewhere "cruciform sovereignty" (Carmody 2009).

A question for globally dominant social forces is how to regulate or reabsorb "surplus populations" (Davis 2006) who have been expelled from the logic of global production and consumption. What I call matrix governance, described in more detail later, attempts to do this through the "discipline" of the market, as the poor in the Global South are encouraged to become entrepreneurs of the self (Foucault 1978). As a discursive strategy this is known as responsibilisation, as the burden of responsibility is discursively placed on the poor themselves for their own poverty and potential upliftment.

A second strategy which is pursued is consumerisation (Murphy and Carmody 2015). For example, mobile telephony is the frontier of consumerisation and "financial inclusion" in sub-Saharan Africa as Facebook, for example, offers free access to a restricted number of websites through its 'basics' package, leading the majority of people in some developing countries, such as Nigeria, to agree with the statement that "Facebook is the internet" (Mirani 2015 cited in Mann 2016). The hard infrastructure of informationalisation, such as fibre-optic cables, is part of the broader infrastructure of matrix governance (Carmody and Surborg 2013). There are historical parallels with earlier periods of technological innovation. As Marx stated in relation to India, "that unity, imposed by the British sword, will now be strengthened and perpetuated by the electric telegraph" (Marx 1853). Now the mobile phone is a vector of neoliberalisation, for example, through so-called fin-tech (financial-technology) products, which will be discussed in more detail later.

Impacts of neoliberalism

> Ending global poverty through economic growth alone will take more than 200 years (based on the World Bank's inhumanly low poverty line of $1.90 a day) and up to 500 years (at a more generous poverty line of $10 a day) . . . . The damage to the natural environment caused by several more hundreds of years of capitalist growth would wipe out any gains in poverty reduction.
>
> – (Selwyn 2017, 3)

Neoliberalism had disastrous economic and social consequences in much of the Global South. In an effort to reinvent it the World Bank and IMF moved away from SAPs to so-called Poverty Reduction Strategy Papers (PRSPs) in 1999. These arguably represent an attempt to counter some of the resistance to neoliberalism in the developing world, in addition to take on learning from the failures of SAPs. SAPs were seen to have underperformed by the World Bank and IMF because they undermined so-called "human capital" formation by cutting back on health and educational expenditures. Human capital theory posits that workers will be more productive if they are educated and healthy, which is correct. However, having a better workforce does not necessarily create jobs for them. Also, some people have questioned the idea of designating people as a form of capital.

PRSPs primarily differed from SAPs through their emphasis on access to universal primary education and primary health care, and consequently the abolition of user fees which had been introduced under SAPs. In some case user fees had been introduced for things such as sexually transmitted disease clinics, with predictable and sometimes fatal consequences – people couldn't afford to attend the clinics and went undiagnosed. These seeming "savings" then resulted in untold suffering and economic losses.

The emphasis on health and education in PRSPs represents what some have referred to as social neoliberalism – an attempt to make it more socially acceptable. However, the economic model behind PRSPs remained substantially unchanged from that of SAPs as they maintained their commitment to "free" trade and sought to raise the profit (capital) share in economic output in order to increase investment by reducing wage costs to make business profitable. Depending on the definition of poverty adopted, some might consider it paradoxical or counterproductive that "poverty reduction" programmes seek to increase inequality. The contradictions of neoliberal "development" have led to other currents and trends in the international development space globally.

Post-neoliberalism and the return of modernisation?

One way to trace changes in development theory and practice is through the personal histories of influential individuals who have played important roles in shaping both of these. One of the most influential people in the world in terms of international development currently is Professor Jeffrey Sachs of Columbia University in New York. As noted earlier, he is special advisor to the United Nations Secretary General on the SDGs and was heavily involved in designing the previous global goals – the MDGs.

Jeffrey Sachs has had a remarkable career, both in terms of his influence and trajectory and his own thinking. He is the subject of at least two books in addition to being a prolific author in his own right. He is noted for being involved in designing and implementing "shock therapy", or dramatic economic liberalisation, in South America and Eastern Europe in the 1990s, and his website correctly notes that "Professor Sachs' work has been pivotal in many of the key junctures of globalization during the past thirty years. In the 1980s he helped several Latin American countries including Bolivia, Brazil, and Peru to end hyperinflations and renegotiate their external debts".

In addition to his role in designing economic reform programmes around the world, he has made other major interventions in the field of international development, including chairing the Commission on Macro-economics and Health for the World Health Organization (WHO), being the driving force behind the Millennium Promise campaign and the Millennium Villages and founding the Sustainable Development Solutions Network, amongst other initiatives. The *New York Times* has described him as "probably the most important economist in the world". Yet he also has attracted controversy and criticism.

One of the most vocal of these critics is the Manchester University lecturer, Japhy Wilson, who writes "Jeffrey Sachs played a central role in the construction of the very same neoliberal system that he is now so vocally opposing" (Wilson 2014, 3). In recent years Jeffrey Sachs has been an outspoken opponent of rising inequality and militarism, addressing the Occupy Wall Street movement in the US, for example, and calling for US troops to be taken out of Afghanistan.

According to Wilson the catastrophic experience of economic reform, in Russia in particular in the 1990s, which resulted in male life expectancy plunging into the fifties and the economy contracting by more than 40 per cent scarred Sachs and resulted in him focussing his attention on the need to increase aid to the world's poorest countries. The Nobel Prize–winning economist Joseph Stiglitz argued in 1999 that the

> failure of reform in Russia was due to "a misunderstanding of the very foundations of a market economy"; "a failure to grasp the fundamentals of reform processes"; and "an excessive reliance on textbook models of economics". Sachs wasn't mentioned by name but he didn't have to be.
>
> (Munk 2013, 17)

The dramatic economic contraction in Russia paved the way for the emergence of Vladimir Putin as a "strong man" ruler for decades. However, some might argue that a dramatic economic restructuring and contraction were inevitable in the shift to a more market-based economy, although perhaps not to the extent that it occurred.

According to Wilson, Professor Sachs has turned Africa into what Slavoj Žižek calls "a sublime object of ideology". However, in this reading Africa does not function just as an object onto which a particular set of ideas about how the world works are projected; rather,

> in its deepest and most powerful form Žižek argues, ideology operates not as an appearance projected onto reality, but as a 'social fantasy' structuring reality against what Lacan called the Real. The Real is an ominous presence that is excluded from everyday experience, but that imposes itself on reality in disturbing and inescapable ways. It is most directly encountered in moments of trauma and psychotic breakdown, in which reality disintegrates and the Real confronts the subject as a terrifying and incomprehensible force. . . . Rather than responding to such traumatic events by discarding their economic model, neoliberals attempt to hold their sense of reality together by explaining their failures in terms that leave their fantasy intact.
>
> (Wilson 2014, 8–9)

It could be argued that in fact Jeffrey Sachs has been quite consistent in his perspectives, however. For example, during the economic reform programme in Russia, he argued for more international aid for the country (Wilson 2014), which was badly needed. Likewise, when he was involved in the economic reform programme in Bolivia, he argued for debt relief for the country and got the IMF to agree to this. This is entirely consistent with his subsequent position and work in Africa. Although he described himself as a free-market ideologue during the reform process in Russia, by the time his primary attention had turned to Africa, he was arguing against "'simplistic free-market ideology'. 'Malaria is not a market', he informed the group of foreign aid donors in Tanzania. 'It's a pandemic disease and a killer. We're not selling Buicks here – we're trying to keep people alive!'" (quoted in Munk 2013, 99).[1] His critique has subsequently gone further still, arguing, for example, that "markets are basically designed to ignore the poor, as they are generally not good consumers" (Sachs 2015, 498).

Perhaps after shock therapy Jeffrey Sach's most controversial development project/intervention has been the Millennium Villages (MVs). These were set up because

> it was realised that most countries in Sub-Saharan Africa were not likely to achieve the [Millennium Development] goals by the year 2015 and thus MVP [the Millennium Villages Project] was born to speed up the achievement of the MDGs. The UN Secretary-General Kofi Annan commissioned the Millennium Project to produce a strategy for the achievement of the Goals, which was then implemented in the Millennium Villages.
>
> (Kimanthis and Hebinck 2016, 1)

The idea behind the MVs was relatively simple and quite intuitive. In his book *The End of Poverty* Professor Sachs (2005) develops the idea of poverty traps. He argues that where places are remote or land-locked, where the burden of disease is high and where people are too poor to save, and therefore investment is low, self-reproducing poverty traps may develop, which is undoubtedly correct, although these traps are transnationally constructed. He also argues in his writing that colonialism was deeply unhelpful for

African development and criticises SAPs for cutting health and education expenditure when more investment in these areas was needed. In order to break these poverty traps, he argued, there was a need to invest in education, infrastructure, agriculture and small business development to move local economies to a new equilibrium, where supply and demand are increased. This is similar to the type of argument made by Nelson (1956) about peasant societies being stuck in what he called low-level equilibria. It could also be seen to be a reinvention of integrated rural development programmes, which sought to leverage synergies between different sectors, such as education, agriculture and infrastructure, to achieve development gains (Oman and Wignaraja 1991), and as such perhaps a return to the idea of modernisation.

Millennium Villages were set up in different countries in Africa in different types of agro-ecologies to show how they could work across contexts (Map 4.1). Some of them were designated as research villages, where the impacts of interventions would be heavily monitored and studied, and they attracted funding from various donors, including the United Nations Development Programme. However, they were controversial for a variety of reasons.

I attended the Millennium Villages Summit in Ethiopia in 2009. At that meeting different villages reported on their progress in, for example, reducing malaria or raising crop yields. The gains were very impressive. For example, in Sauri in Kenya, maize production tripled as a result of the provision of inputs such as fertiliser, and the prevalence of malaria fell from 55 to 13 per cent (Millennium Promise 2008; See also Nziguheba *et al.* 2010). Other work has also shown that Millennium Villages have dramatically improved access to improved water sources in Senegal, for example (Murphy-Teixidor 2018).

The idea behind the MVs was that the equilibrium could be shifted from a vicious to a virtuous cycle, so that, for example, as people's health improved, they would be more productive. Jeffrey Sachs claimed that the Millennium Villages were like an aeroplane – if they didn't gain enough momentum, they would never take off (Munk 2013); consequently, sufficient funding was needed. The Millennium Villages provided much-needed health clinics, for example, and, in one of the villages I visited, access to the Internet through a weather-proof computer station embedded in a wall, which, however, did not seem to be receiving much use.

One of the issues which was raised by one of the project managers at the summit I attended was that the results of the programme were being diluted by in-migration by people in surrounding districts to access health and other services. He suggested that if there was a way to reduce this through the establishment of buffer zones around the villages, this might be beneficial. However, such a policy was correctly not implemented. If it had been, it would have created what one overseas development assistance official referred to as "fortress Millennium Villages". This same official noted to me that it was not surprising that the MVs produced results, as they had considerable investment put into them. However, this official was sceptical about their scalability, although the Nigerian government adopted the model as official government policy. The issue was perhaps one more of funding than inherent lack of scalability, or how scalability is defined.

The resource issue was a controversial one in the MVs. In one of them a sign was put up that visitors should not give things or money to the locals, as they wanted to maintain

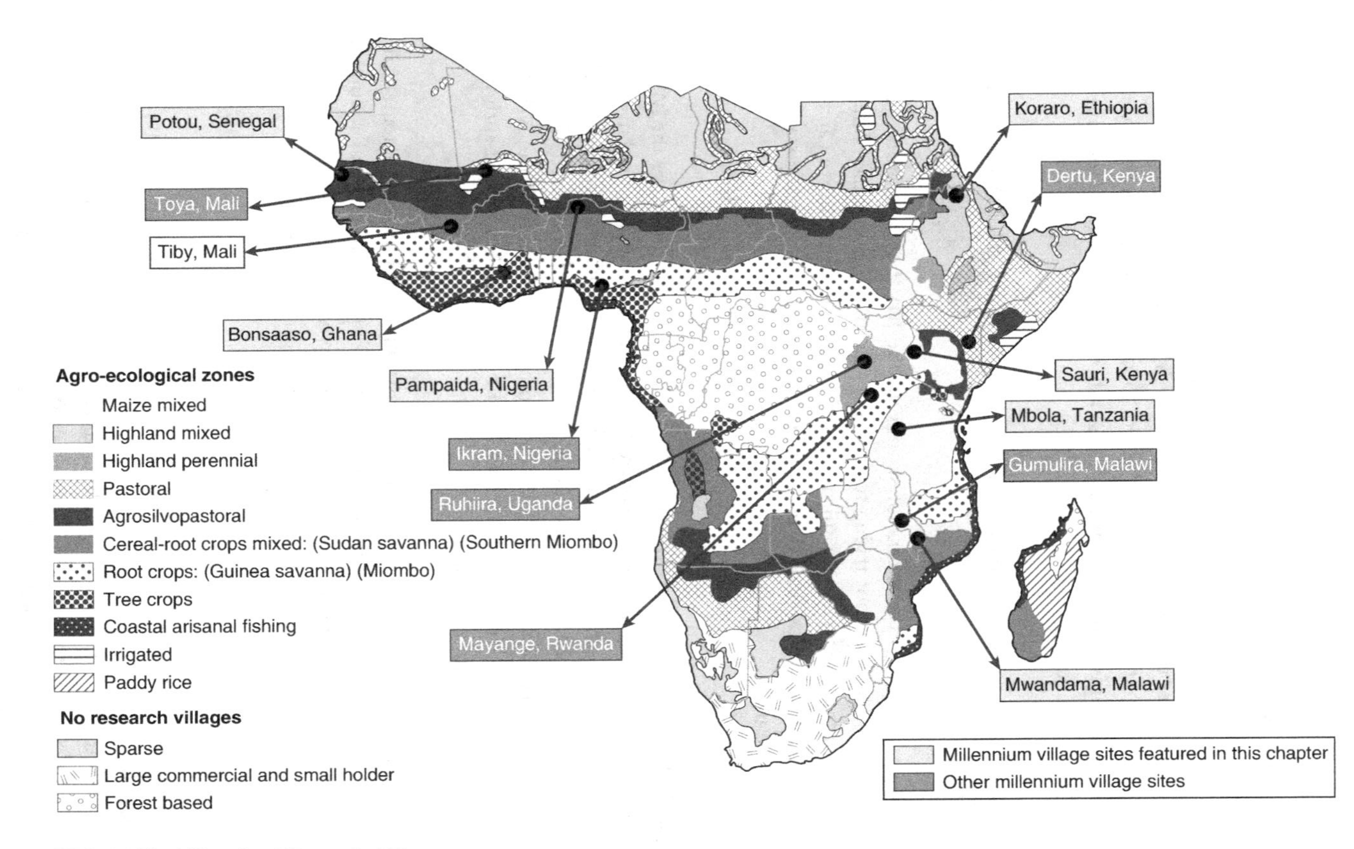

MAP 4.1 The Millennium Villages in Africa

Source: Reproduced from the Sanchez, Palm, Sachs *et al.* (2007) by permission. Copyright (2007) National Academy of Sciences, USA.

the integrity of the research project (Wilson 2014). One commentator drew a comparison with the "don't feed the animals" signs at zoos. The amount of per capita investment in the MVs was meant to be modest – about US $120 per person per year (Tollefson 2015) – to show that the impacts of modestly increased aid (per capita) could have significant results and consequently be scalable. However, Michael Clemens (2012, 1) of the Centre for Global Development found that

> the project costs at least US $12,000 per household that it lifts from poverty – about 34 times the annual incomes of those households . . . the same $12,000 in a bank account at 5% interest would yield $600, every year, year after year, forever. That interest, given to the households as cash, would cause their incomes to nearly triple, permanently.

There was also some controversy around the ways in which the MVs were monitored and evaluated:

> Sachs has faced criticism for not setting up the MVP as a rigorous experiment. MVP researchers are not trying to retroactively to compare villages that received the full intervention with similar ones that did not, but the research protocol readily acknowledges challenges in collecting data and producing statistically significant results.
>
> (Tollefson 2015, 144)

Although this criticism is perhaps somewhat unfair, as it was only after the MVP project was operational that randomised controlled trials became widely used in economics and these have many of their own problems, as discussed later. Professor Sachs also addressed criticism around evaluation in the video referenced at the end of this chapter.

Some of the claims made in a published internal evaluation of the MVP in the prestigious medical journal, *The Lancet*, had to be retracted. As the lead author of that evaluation wrote in the retraction,

> we made some erroneous statements and assumptions. First, the statement in the Findings section of the Summary that 'the average annual rate of reduction of mortality in children younger than 5 years of age was three-times faster in Millennium Village sites than in the most recent 10-year national rural trends (7.8% vs. 2.6%)' is unwarranted and misleading.
>
> (Pronyk 2012)

Importantly, however, the editors of the journal noted that the main findings from the paper stood.

There were also some other issues identified with evaluation of the MVs, perhaps unsurprisingly, given the scale and ambition of the programme. For example, in its 2008 annual report the MVP stated that the livestock market that had been set up in Dertu, another MV in Kenya, under its auspices was generating a profit of US$14,000 a month. However, according to Munk (2013, 160–1), "In reality the livestock market never generated a cent . . . the whole concept of selling one's livestock is antithetical to Somali values". The village dispensary had also been stocked with contraceptives, but only five or six people in the village used birth control.

The MVP was also criticised for its conceptualisation of rural African life as characterised by undifferentiated and poor peasant farmers who need to adopt modern production techniques. Rather, some have argued, Africa rural societies are highly class-stratified, and it was often wealthier farmers and land owners who were best able to access the benefits of the MVs (Wilson 2014). Furthermore, the performance of the MVs is influenced by such regional politico-economic conditions. It was claimed that the MV in Ghana was very successful, even as it was clear that socio-economic conditions were disimproving as a result of a gold rush in the region, which displayed the symptoms of the "resource curse" (Wilson 2016). Ultimately, the MV had to be moved to a new site in the country. Although one case would not necessarily invalidate the project, and there is an increasing recognition in the development literature that things not going to plan is inevitable and generates important learnings. Indeed, there is now a whole literature on "iterative adaptation" in development, which essentially means adjusting projects and programmes to work better as they are implemented and conditions evolve (see for example Andrews *et al.* 2013).

Professor Sachs's conceptualisation of poverty has also been critiqued, with some arguing that it is impossible to eliminate poverty because capitalism is dependent on the production of inequality, and consequently relative poverty, for its continued existence (Unwin 2007). However, Professor Sachs is correct that it should be possible to eliminate extreme or absolute poverty through mechanisms such as international redistribution of income, debt relief, fairer trading rules and regimes, governance reform and other mechanisms. The primary problem is that current arrangements benefit the world's principal power holders, and they are invested in their perpetuation. Climate change, through time, may, however, cause them to rethink their positions.

Jeffrey Sachs's work on climate change mitigation is brilliantly incisive and visionary (Sachs and Someshwar 2012). He is the world's leading development economist on many accounts and has made major contributions to the field, in addition to being highly controversial. In many ways he is an organic intellectual of the era of globalised capitalism, working both with multi-national corporations while also claiming allegiance to the 99 per cent, as opposed to the 1 per cent (Wilson 2014). He is critical of "greedy people", while also having to work with them – perhaps out of perceived necessity if the global system is to become more progressive. He has also been a strident public voice against Donald Trump's policies and politics and advocates for a more progressive American foreign policy (Sachs 2018).

Celebratisation of development: normalising neoliberalism or pragmatic power?

Jeffrey Sachs is a visionary leader and inspirational speaker; however, another way that he has been controversial is that he has arguably facilitated the "celebratisation of aid" through work with Bono and Angelina Jolie, for example. Bono wrote the foreword to

Jeffrey Sachs's book on *The End of Poverty*, and they have also travelled and campaigned together, as Jeffrey Sachs has also with Angelina Jolie. An African activist quoted in Moyo (2009) noted that their voices couldn't compete with an electric guitar. Probably the people most associated with international development in the Western public mind are people like Bono, Bob Geldof and Angelina Jolie, and yet nobody has elected them to speak for African or other publics, for example. It was Bob Geldof, the Irish rock singer, businessman and organiser of the Live Aid rock concerts in the 1980s, who convinced the British prime minister Tony Blair to set up the Commission for Africa (Commission for Africa 2005). Bono, the lead singer of the rock group U2, set up the One Foundation to lobby for increased aid and debt relief. He also founded Product Red, which is where companies producing cell phones, for example, donate part of their profits from designated branded products to fund AIDS relief by giving money to the UN Fund for AIDS, Tuberculosis and Malaria. This initiative has been heavily criticised by some scholars as promoting inappropriate conceptions of Africans as powerless, for example, and the idea that consumerism, or causumerism, can solve the world's problems (Richey and Ponte 2011).

Bono has also been the subject of multiple biographies, including *The Frontman: Bono (in the Name of Power)* (Browne 2013). In this book Browne critiques everything from Bono's politics to his song lyrics and is particularly critical of what he identifies as celebrity humanitarianism. He quotes Audrey Bryan (2012, 276) that

> while ostensibly *about* the lives of those whom they seek to uplift and save, discourses of high profile Western benevolence, concern and compassion actively position "our guys" as *the* stars of the development show, while the objects of national (and Northern) benevolence merely function as the backdrop to a story which is really about "us". In other words, the trope of celebrity humanitarianism functions as a redemption fantasy . . . wherein inhabitants of the Global South are discursively positioned as the backdrop against which the "global good guys" can enhance their sense of themselves, and the reputation of the nation they represent, with insufficient attention to their own participation in relations of domination. (79)

However, Bono does recognise the potentially destructive effects of global capitalism, saying that, perhaps echoing Amory Lovins, "capitalism is not immoral, but it is amoral. And it requires our instruction. It's a wild beast that needs to be tamed, a better servant than master" (Bono n.d.). This would imply support for strongly state-regulated or planned capitalism, although in practice Bono and the company that manages U2's affairs have taken advantage of possibilities for tax avoidance through moving its headquarters to the Netherlands. Clement Attlee, the British Labour Party prime minister after World War II, argued that "charity is a cold, grey, loveless thing. If a rich man wants to help the poor, he should pay his taxes gladly, not dole out money at a whim" – a sentiment echoed by Jeffrey Sachs in his book on *The Price of Civilization* (2011). However, there is a lively debate about the nature and impacts of philanthropy and what has been called philanthrocapitalism, which runs more on business principles, particularly funds for charity being drawn from investment revenue. However, Bono has had a generally very positive impact through his campaigning for debt relief, even if he is one of the "have yachts", rather than "have nots", as he said at a British Labour Party conference.

ODA or aid will be discussed in more detail later, but in relation to the ethics of foreign aid there are two main approaches: deontological, or duty-based, ones and consequentialist, or outcomes-based, ones (Murphy 2016). Ontology refers to what exists, so a deontological approach could be said to refer to what should exist. On the other hand, consequentialists argue that actions should be judged by their outcomes. This is a useful way to think about celebrity humanitarianism. There are certainly issues surrounding representation – for example, Bono made a major contribution to the Make Poverty History Campaign, which contributed to securing very substantial debt relief for many African countries through the Multilateral Debt Relief Initiative, which has had very substantial positive impacts (Radelet 2010). Thus, from a consequentialist perspective Bono has arguably had a very positive impact on international development, even if he still celebrates Band Aid lyrics, which talk about there not being snow in Africa and asks whether people in Africa know it is Christmas (Miller 2015). This could be read as patronising, implying ignorance. In any event why should non-Christians care or know when Christmas is? The song implies a need to educate and perhaps convert or evangelise people in Africa. In other cases, though, celebrity interventions have produced unintended and sometimes negative consequences.

Coltan is a semi-precious metal which works as an electrical capacitator in electronic devices. It's extraction in the DRC has been associated with extreme (sexual) violence and exploitation (Nest *et al.* 2006). As part of the Dodd-Frank Act in the United States, introduced to better regulate the financial sector after the NAFC, imports of "conflict coltan" from the DRC were banned. This is a relatively common practice in US law making, where seemingly unrelated issues are attached to bills that are likely to pass. The American actor Ben Affleck had lobbied hard for this inclusion (Chase 2015). However, while the motivation behind this was well intentioned, it resulted in perhaps millions of miners in the DRC losing their livelihoods when imports of coltan from the region were dramatically reduced (Wolfe 2015). While one of the goals of the measure had been to reduce sexual violence, the increased immiseration of miners and communities did not promote this.

The SDGs – socialising neoliberalism or global partnership for development?

The SDGs were adopted by the United Nations in December 2015. They replace the MDGs, which had been adopted in 2000 and expired in 2015. There are a number of features which distinguish the SDGs from the MDGs. The first of these is that whereas the MDGs were focussed on developing countries, the SDGs are global, given the recognition that with globalisation, including of the environment, what happens in one place affects others. The second is the number of goals – the MDGs had eight, whereas the SDGs have seventeen. In part this reflects the way in which the different sets of goals were negotiated. Whereas the MDGs were designed by a small project team, the SDGs were more political and involved inter-governmental negotiation through an "open working group" (OWG), which used a system of representation where most of the seats were shared by several countries. The OWG was co-chaired by the Kenyan and Irish ambassadors to the United Nations.

The fact that the process around the development of the SDGs was more inclusive and largely inter-governmental in nature could be one of the reasons that there were more goals than in the case of the MDGs. The MDGs' design appears to have been somewhat ad hoc. Mark Malloch-Brown, who was then the administrator or head of the United Nations Development Programme, recounts that

> the document had gone to the printing presses as I passed the head of the UN's environmental programme . . . I was walking along the corridor, relieved at job done, when I ran into the beaming head of the UN environment programme and a terrible swearword crossed my mind when I realised we'd forgotten an environmental goal . . . we raced back to put in the sustainable development goal.
>
> (quoted in Tran 2012)

There was widespread dissatisfaction with the way the MDGs were negotiated. For example, the well-known Egyptian-Franco political economist, Samir Amin (2006), argued that they showed the largely untrammelled power of the US after the collapse of the Soviet bloc, including allowing a consultant for the CIA (Ted Gordon) to be involved in drafting them. However, for others, they did broaden the mainstream development agenda beyond a focus on economic growth (Vandemoortele and Delamonica 2010 cited in Mutasa 2015), even if they did not challenge neoliberalisation. Interestingly, however, the SDGs are more overt in their promotion of the market, with Goal 17 calling for countries to "adopt and implement investment promotion regimes for LDCs [Less Developed Countries]" to promote foreign investment (quoted in Brooks 2017, 203). Likewise the Addis Ababa Action Agenda for financing for development, which was adopted by the UN General Assembly and fed into the SDGs, said that

> [i]nternational trade is an engine for inclusive economic growth and poverty reduction, and contributes to the promotion of sustainable development. We will continue to promote a universal, rules-based, open, transparent, predictable, inclusive, non-discriminatory and equitable multilateral trading system under the World Trade Organization (WTO), as well as meaningful trade liberalization.
>
> (quoted in Sustainable Development Knowledge Platform 2015)

The SDGs run for a period of 15 years, from 2016 to 2030 (hence what is referred to as "Agenda 2030"). However, while they were adopted in 2000 and finished in 2015, the MDGs actually ran, in theory, for 25 years, as the baseline for poverty reduction and the other goals was set at 1990. This had the advantage that all of the massive poverty reduction that took place in China during that 25-year period could be captured – meaning that the target of halving global (absolute) poverty was actually met five years ahead of schedule (United Nations n.d.).

The global focus on the SDGs has a number of implications in terms of delivery, shifting the emphasis away from aid as the modality of assistance in the MDGs, which had the effect of reinforcing "vertical interventions" from the Global North to the South (Fehling *et al.* 2013), to mainstreaming across all government departments in all members of the UN. Mainstreaming is a method of policy implementation common in the new public management theory, which has a variety of aspects or dimensions but seeks to apply business principles to government. An example of the way this works is that rather than just having policies to promote gender equity, this could be mainstreamed across all government departments so that all government policies would try to promote this. However, neither the MDGs nor SDGs specify particular accountable parties beyond governments in general. As such, they are somewhat aspirational – a form of "soft law" – and seem to operate on a similar basis to the European Union's Open Method of Coordination, where countries voluntarily cooperate to achieve common aims or goals but determine themselves how to achieve these. As such, if goals or targets within them are not met, there are no repercussions beyond perhaps those of negative publicity and "audience effects" for domestic citizens, who could vote governments who fail to meet them out of power. Africa, for example, missed many of the MDG goals (Kankwenda 2015). In part this is because of the conditionalities attached to IFI assistance and aid more generally. The MDGs and SDGs partly arose out of an increased interest in "human development", which is a concept most associated with the work of the Nobel Prize–winning economist Amartya Sen.

The capabilities or human development approach is most associated with Amartya Sen, and particularly his book *Development as Freedom* (Sen 1999). In this book he sets out the case as to why the expansion of freedom should be both the primary means and end of development. This is often seen in contrast to conventional economic theory, which prioritises economic growth. However, it could be argued that conventional economics seeks to maximise utility for society and that utility or happiness is a form of freedom, so the distinction or disjuncture might not be as sharp or defined as is often thought. In his book Sen seeks to set out how people can lead lives they value, free of the social shame of poverty. They can do this, he argues, by being facilitated in maximising capabilities or abilities and functionings or choices (Selwyn 2014).

At first, the capability approach appears to achieve a reconciliation between palliative and structural conceptions of poverty, with Sen explicitly stating that his framework draws on the work of both Karl Marx and Adam Smith (Clark 2006). Palliative views of poverty see it as residual – something to be "mopped up" because there are specific market failings. On the other hand, more structural approaches view poverty as something which is produced through the operation of the market. However, the unit of analysis of the capability approach is the individual (Hill 2007), and this obscures issues of class power and, in particular, the class nature of the state (Jessop 2002), which is charged with implementing policies to overcome poverty. "Following Adam Smith, Sen acknowledges that capitalist markets are a source of economic growth, but his principle argument is that they are sources of individual freedom" (Selwyn 2014, 166). In this he shares similarities with neoliberal conceptions of development, which have arguably heightened many inequalities around the world, including those around gender.

Neoliberalism, gender and development

It is often said that poverty has a woman's face, as it is an oft-cited "fact" that 70 per cent of the people around the world living in absolute poverty are women. The source for this claim is the 1995 Human Development Report of the United Nations Development Programme, but no detail is provided as to how this figure was arrived at, and in some parts of the report it is said to refer to women but in other places to women and girls (Green 2008). However, as with many such accepted "facts", the actual position is more complex.

In the first instance data on poverty are often collected at the household level, disguising gender disparities. "Simple disaggregation of poverty by sex without taking into account intrahousehold inequality results in small but probably underestimated poverty gaps" (United Nations Department of Economic and Social Affairs 2010, 158). In some cases, the results are counter-intuitive as "only in 4 of the 16 countries in Africa with available data – Burundi, Malawi, Sao Tome and Principe and Zambia – were the poverty rates for female-headed households higher compared to male-headed households" (161). However, this does not mean that there are not substantial gender differentials within households, which also affect access to public services.

Literacy levels are an extreme example of this. In 2009 in sub-Saharan Africa 74.8 per cent of men were literate, whereas the equivalent figure for women was only 56.3 per cent (World Bank 2011). In Chad less than a quarter of women are literate. This obviously affects not only quality of life but also lifetime earnings. In some cases gender disparity is even more pronounced through access to life itself. Around the world 106 male children are born for every 100 female children (Hassan 2014). However, men are less biologically resilient and more likely to die at every age, all other things being equal (Saini 2017). In 1990 Amartya Sen (1990) estimated that there were more than 100 million women who should have been alive at that time but weren't – many of them in India. This relates to practices such as gender-selective abortion, infanticide (where children are murdered) and unequal treatment between boys and girls in terms of being taken to the doctor when ill or being given sufficient food. Male preference or bias is often related to broader issues of male dominance in society, also known as patriarchy.

There is immense variation around the world in terms of gender – that is socially constructed norms around roles and behaviour associated with biological sex – but given the prevalence of patriarchy around the world, women often tend to serve as "shock absorbers" for social or economic problems (Shiva 1988). Sometimes unequal treatment of women is thought to be associated with "tradition"; however, European colonialism often brought gender disempowerment with it as women were sometimes forbidden from owning land or dethroned as monarchs. Queen Nzinga in what is modern-day Angola spent much of her life fighting Portuguese occupation (Miller 1975).[2] In other cases gender discrimination has been associated with technological development. For example, the so-called "Green Revolution" in India, which depended on increased use of mechanised technology, devalued women's labour and thereby increased the number of "missing women" by increasing female infanticide (Shiva 1988).

Earlier approaches such as "women in development" have been critiqued for arguing that if modernisation was undertaken properly, it could benefit women (Prügl 1999).

However, the interactions between neoliberalism and gender are complex, depending on what geographers call site and situational characteristics. As noted earlier, site characteristics are those internal to a place, whereas situation refers to the nature of interactions between places. In some places it appears that free-market reforms created jobs for women in much of Southeast Asia, for example, although the quality of these has often been poor. In other instances of industrial restructuring women were often the first to lose their jobs in the textile industry in Nigeria, for example, as the country was flooded by Chinese imports (Burgis 2015).

Often women's work in the family or home goes unrecorded in the national statistics. This has led some to argue that capitalism is parasitic on unpaid women's work to reproduce itself (Mies 1986). To the extent that women are more reliant on public services for childbirth, for example, or to offset the socially constructed economic advantages men have in much of the world, they were often more severely affected by policies of austerity under structural adjustment. This is one of the channels through which the feminisation of poverty has operated. However, as Chant (2008, 165) argues,

> foremost among my conclusions is that since the main indications of feminisation relate to women's mounting responsibilities and obligations in household survival we need to re-orient the 'feminisation of poverty' thesis so that it better reflects inputs as well as incomes, and emphasises not only women's level or share of poverty but the burden of dealing with it.

Taking this perspective, the feminisation of poverty would appear to be associated with neoliberalism. According to UN Women (n.d.):

> Women in most countries earn on average only 60 to 75 per cent of men's wages. Contributing factors include the fact that women are more likely to be wage workers and unpaid family workers; that women are more likely to engage in low-productivity activities and to work in the informal sector, with less mobility to the formal sector than men; the view of women as economic dependents; and the likelihood that women are in unorganized sectors or not represented in unions.

Some have argued that oppressive labour regimes for women workers in export-processing zones, for example, are parasitic on patriarchy. Mass faintings by women workers are very common in "global factories" in response to the extreme pressures of work there. The "super-exploitation", or working excessively long hours in oppressive conditions, of women in factories was an important part of the growth of the Asian "tiger" economies discussed in more detail in the next chapter.

Asia has been the world's most developmentally successful region in the last several decades. The original "four little dragons" or "Asian tiger" newly industrialised countries of Singapore, Hong Kong, South Korea and Taiwan have in more recent times been joined by other fast-growing economies in the region, particularly China. Did these countries develop on the basis of "free market" policies, as some claim, or have the state and global factors played important roles? The next chapter will engage and discuss these issues.

Notes

1 According to Professor Sachs, the word "malaria" comes from the Italian words for bad air (*malaria*), as they were convinced that it was spread through the air. The fact that Italy and Florida were able to eliminate the spread of malaria through investments and public health interventions shows that malarious regions are not condemned to permanently suffer from the disease, but that it is both an outcome and cause of poverty.

2 Reportedly the origin of the country's name was that when the Portuguese came they asked the name of the territory and were told it was called "Ngola", which was actually the term for king.

Further reading

Books

Bishop, M., and M. Green. 2008. *Philanthrocapitalism: How the Rich Can Save the World and Why We Should Let Them*. London: A&C Black.

Glennie, J. 2008. *The Trouble with Aid: Why Less Could Mean More for Africa*. London: Zed Books.

McGoey, L. 2016. *No Such Thing as a Free Gift: The Gates Foundation and the Price of Philanthropy*. London: Verso.

Onimode, B. 1989. *The IMF, the World Bank and African Debt*. London: Zed Books.

Articles

Denning, G., P. Kabambe, P. Sanchez, A. Malik, R. Flor, R. Harawa, P. Nkhoma, C. Zamba, C. Banda, C. Magombo, M. Keating, J. Wangila, and J. Sachs. 2009. "Input Subsidies to Improve Smallholder Maize Productivity in Malawi: Toward an African Green Revolution." *Plos Biology* 7 (1): 2–10.

Wilson, J. 2016. "The Village that Turned to Gold: A Parable of Philanthrocapitalism." *Development and Change* 47 (1): 3–28.

Websites

The African Forum and Network on Debt and Development, www.afrodad.org

Ethics Matter: A Conversation with Jeffrey D. Sachs, Carnegie Council, www.youtube.com/watch?v=sQGsJAPjLL0

Millennium Promise, www.millenniumpromise.org/en/about-us

Millennium Villages, http://millenniumvillages.org/

Sustainable Development Solutions Network, http://unsdsn.org/

References

Andrews, M, L. Pritchett, and M. Woolcock. 2013. "Escaping Capability Traps Through Problem-Driven Iterative Adaptation in Development." *World Development* 51: 234–44.

Amin, S. 2006. "The Millennium Development Goals: A Critique from the South." *Monthly Review-an Independent Socialist Magazine* 57 (10): 1–15.

Brooks, A. 2017. *The End of Development: A Global History of Poverty and Prosperity*. London: Zed Books.

Browne, H. 2013. *The Frontman: Bono (In the Name of Power)*. London: Verso.

Bryan, A. 2012. "Band-Aid Pedagogy, Celebrity Humanitarianism, and Cosmopolitan Provincialism: A Critical Analysis of Global Citizenship Education." In *Ethical Models and Applications to Global-*

ization: Cultural, Socio-Political and Economic Perspectives, eds. C. Wankel and S. Malleck. Hershey, PA: Business Science Reference.

Burgis, T. 2015. *The Looting Machine: Warlords, Tycoons, Smugglers, and the Systematic Theft of Africa's Wealth.* London: William Collins.

Carmody, P. 2009. "Cruciform Sovereignty, Matrix Governance and the Scramble for Africa's Oil: Insights from Chad and Sudan." *Political Geography* 28 (6): 353–61.

———. 2015. "Ecolonisation and the Creation of Insecurity Regimes: Zimbabwe's Land Reform in Regional Context." In *State, Land and Democracy in Southern Africa*, eds. C. Tornimbeni and A. Palloti. Burlington, VT and London: Ashgate.

———. 2016. *The New Scramble for Africa*. 2nd edition. Cambridge: Polity.

———., and B. Surborg. 2013. "Of Cables, Connections and Control: Africa's Double Dependency in the Information Age." In *Enacting Globalization: Multidisciplinary Perspectives on International Integration*, ed. L. Brennan. Basingstoke and New York Palgrave Macmillan.

Chant, S. 2008. "The 'Feminisation of Poverty' and the 'Feminisation' of Anti-Poverty Programmes: Room for Revision?" *Journal of Development Studies* 44 (2): 165–97.

Chase, S. 2015. "We Will Win Peace." www.kanopy.com/product/we-will-win-peace.

Clark, D. 2006. "The Capability Approach: Its Development, Critiques and Recent Advances." http://economics.ouls.ox.ac.uk/14051/1/gprg-wps-032.pdf.

Clemens, M. 2012. "New Documents Reveal the Cost of 'Ending Poverty' in a Millennium Village: At Least $12,000 Per Household." www.cgdev.org/blog/new-documents-reveal-cost-"ending-poverty"-millennium-village-least-12000-household.

Commission for Africa. 2005. *Our Common Interest: The Commission for Africa: An Argument.* London: Penguin Books.

Davis, M. 2006. *Planet of Slums.* London and New York: Verso.

Fehling, M., B.D. Nelson, and S. Venkatapuram. 2013. "Limitations of the Millennium Development Goals: A Literature Review." *Global Public Health* 8 (10): 1109–22.

Foucault, M. 1978. "The Birth of Bio-Politics." *Lecture at the Collège de France.* https://www.thing.net/~rdom/ucsd/biopolitics/NeoliberalGovermentality.pdf.

Green, D. 2008. *From Poverty to Power: How Active Citizens and Effective States Can Change the World.* Oxford: Oxfam Publishing.

Harrison, G. 2004. *The World Bank and Africa: The Construction of Governance States.* London: Routledge.

Hassan, R. 2014. "The 'Missing Women' in India." *Institute of South Asian Studies Working Paper No. 195, National University of Singapore.* www.files.ethz.ch/isn/184037/ISAS_Working_Paper_No__195_-_The_'Missing_Women'_in_India_19092014174104.pdf.

Hill, M. 2007. "Confronting Power Through Policy: On the Creation and Spread of Liberating Knowledge." *Journal of Human Development* 8 (2). www.capabilityapproach.com/pubs/Hill07.pdf.

Jessop, B. 2002. *The Future of the Capitalist State.* Cambridge: Polity.

Kankwenda, M. 2015. "Rethinking the Vision for Development in Africa." In *Africa and the Millennium Development Goals: Progress, Problems, and Prospects*, eds. C. Mutasa and M. Paterson. London: Rowan and Littlefield.

Kautsky, K. 1914. "Der Imperialismus." *Die Neue Zeit* 2: 908–22.

Kimanthis, H., and P. Hebinck. 2016. "Castle in the Sky: Sauri Millennium Village in Reality." In *Global Governance/Politics, Climate Justice & Agrarian/Social Justice: Linkages and Challenges.* The Hague: Institute for Social Studies.

Mann, L. 2016. "Corporations Left to Other People's Devices: A Political Economy Perspective on the Big Data Revolution in Development." *Development and Change* 49 (1): 3–36.

Marx, K. 1853. "The Future of British Rule in India." *New-York Daily Tribune*, August 8th.

Mies, M. 1986. *Patriarchy and Accumulation on a World Scale: Women in the International Division of Labour.* London: Zed Books.

Millennium Promise. 2008. "Baseline Report of the MDG Indicators, Sauri, Kenya." www.millenniumpromise.org/site/PageServer?pagename=mv_1sauri.

Miller, J.C. 1975. "Nzinga of Matamba in a New Perspective." *Journal of African History* 16 (2): 201–16.

Miller, M. 2015. "Poverty, Inc." Roco Films.

Moyo, D. 2009. *Dead Aid: Why Aid Is not Working and How There Is Another Way for Africa*. London: Allen Lane.

Mountz, A. 2011. "The Enforcement Archipelago: Detention, Haunting, and Asylum on Islands." *Political Geography* 30 (3): 118–28.

Munk, N. 2013. *The Idealist: Jeffrey Sachs and the Quest to End Poverty*. New York: Doubleday.

Murphy Teixidor, A.M. 2018. *An Integrated Approach to Water Security Assessment in Senegal: Metrics Beyond the MDGs and SDGs*. Paper presented at the Annual Development Studies Association of Ireland Conference, Dublin. October 24th.

Murphy, J.T., and P. Carmody. 2015. *Africa's Information Revolution: Technical Regimes and Production Networks in South Africa and Tanzania*. Chichester, West Sussex, UK and Malden, MA: John Wiley & Sons Inc.

Murphy, S. 2016. *Responsibility in and Interconnected World: International Assistance, Duty and Action*. Switzerland: Springer.

Mutasa, C. 2015. "Introduction." In *Africa and the Millennium Development Goals: Progress, Problems, and Prospects*, eds. C. Mutasa and M. Paterson. London: Rowman and Littlefield.

Nelson, R. 1956. "A Theory of the Low-Level Equilibrium Trap in Underdeveloped Economies." *American Economic Review* 46 (5): 894–908.

Nest, M., F. Grignon, and E. Kisangani. 2006. *The Democratic Republic of Congo: Economic Dimensions of War and Peace*. Boulder, CO and London: Lynne Rienner Publishers.

Nziguheba, G. et al. 2010. "The African Green Revolution: Results from the Millennium Villages Project." *Advances in Agronomy* 109: 75–115.

Oman, C., and G. Wignaraja. 1991. *The Postwar Evolution of Development Thinking*. London: Macmillan in Association with the OECD Development Centre.

Pronyk, P. 2012. "Errors in a Paper on the Millennium Villages Project." *Lancet* 379 (9830): 1946.

Prügl, E. 1999. *The Global Construction of Gender: Home-Based Work in the Political Economy of the 20th Century*. New York: Columbia University Press.

Radelet, S. 2010. *Emerging Africa: How 17 Countries Are Leading the Way*. Baltimore, MD: Center for Global Development.

Richey, L., and S. Ponte. 2011. *Brand Aid: Shopping Well to Save the World*. Minneapolis, MN: University of Minnesota Press.

Sachs, J. 2005. *The End of Poverty: How We Can Make It Happen in Our Lifetime*. London: Penguin Books.

———. 2015. *The Age of Sustainable Development*. New York: Columbia University Press.

———. 2018. *A New Foreign Policy: Beyond American Exceptionalism*. New York: Columbia University Press.

———. and S. Someshwar. 2012. *Green Growth and Equity in the Context of Climate Change: Some Considerations*. Asian Development Bank Institute.

Saini, A. 2017. "The Weaker Sex? Science That Shows Women are Stronger Than Men." *The Guardian*. www.theguardian.com/world/2017/jun/11/the-weaker-sex-science-that-shows-women-are-stronger-than-men.

Sanchez, P., et al. 2007. "The African Millennium Villages." *Proceedings of the National Academy of Sciences*, 104 (43): 16775–80.

Selwyn, B. 2014. *The Global Develoment Crisis*. Cambridge: Polity.

———. 2017. *The Struggle for Development*. Cambridge and Malden, MA: Polity.

Sen, A. 1990. "More Than 100 Million Women Are Missing." www.nybooks.com/articles/1990/12/20/more-than-100-million-women-are-missing/.

———. 1999. *Development as Freedom*. Oxford: Oxford University Press.

Shiva, V. 1988. *Staying Alive: Women, Ecology and Development*. London: Zed Books.

Sustainable Development Knowledge Platform. 2015. "A/CONF.227/L.1 – Trade." https://sustainabledevelopment.un.org/index.php?page=view&type=2002&nr=247&menu=35.

Tollefson, J. 2015. "Flagship Aid Programme Up for Evaluation." *Nature* 524: 144–5.

Tran, M. 2012. "Mark Malloch-Brown: Developing the MDGs Was a Bit Like Nuclear Fusion." www.theguardian.com/global-development/2012/nov/16/mark-malloch-brown-mdgs-nuclear.

UN Women. n.d. "Facts and Figures: Economic Empowerment: Benefits of Economic Empowerment" http://www.unwomen.org/en/what-we-do/economic-empowerment/facts-and-figures.
United Nations Department of Economic and Social Affairs, Statistics Division. 2010. "The World's Women 2010: Trends and Statistics." https://unstats.un.org/unsd/Demographic/Products/Worldswomen/WW2010pub.htm#.
Unwin. T. 2007. "No End to Poverty." *The Journal of Development Studies* 43 (5): 929–53.
Urry, J. 2014. *Offshoring*. Cambridge: Polity.
Wilson, J. 2014. *Jeffrey Sachs: The Strange Case of Dr Shock and Mr Aid*. London: Pluto.
———. 2016. "The Village that Turned to Gold: A Parable of Philanthrocapitalism." *Development and Change* 47 (1): 3–28.
Wolfe, L. 2015. "How Dodd-Frank Is Failing Congo." *Foreign Policy*. http://foreignpolicy.com/2015/02/02/how-dodd-frank-is-failing-congo-mining-conflict-minerals/.
World Bank. 2011. *African Development Indicators*. Washington, DC: World Bank.

CHAPTER

5

The role of the state

"Developmental states", geopolitics, industrialisation and security

> The hidden hand of the market will never work without a hidden fist – McDonald's cannot flourish without McDonnell Douglas, the builder of the F-15 [fighter aircraft]. And the hidden fist that keeps the world safe for Silicon Valley's technologies is called the United States Army, Air Force, Navy and Marine Corps.
>
> – (Friedmann 1999)

States

There are a variety of definitions of what constitutes a "state" and also many theories about their purpose and nature. For Strange (1996), states are institutions to mediate and resolve conflicts. However, this arguably neglects how states initially, and the inter-state system more generally, came into being, often through war (for a discussion of the emergence of the inter-state system see Spruyt 1996). On the other hand, for Jessop (2002), states are social relations. For him the state is "an ensemble of socially embedded, socially regularized and strategically selective institutions, organizations, social forces and activities organized around (or at least selectively involved in) making collectively binding decisions for an imagined political community" (6). This terminology of an imagined political community draws on Benedict Anderson's work (1983). Anderson argued that it was the printing press which was largely responsible for the spread of nationalism, by exposing people to the same ideas and constructing shared ideas of history and culture. As we will never meet all of the people living in our countries, they are "imagined" rather than actual communities. The extent to which this common imaginary applies to "weak" or "failed" or "quasi" (Jackson and Rosberg 1982) states in parts of the Third World can, however, be debated. Using terminology like this has been criticised for encouraging Western intervention in the developing world. For example, according to the radical geographer, Richard Peet, if a state is "failed", it can be argued that the "normal" rules of non-interference in internal affairs should not apply, as the state does not exercise territorial sovereignty effectively.

States serve, or are often meant to serve, a variety of functions, including the provision of security and public goods, management of the economy and the reduction of inequalities generated by and through markets. However, different states have different emphases, even if most states around the world are now capitalist in the sense that they generally serve to promote economic growth and capital accumulation for reasons of legitimation of the existing social order and distribution of property, perceived collective good and sometimes the self-interest of office holders.

States and development

For mainstream analysts development is synonymous with economic growth, industrialisation and modernisation. However, as noted previously, the term "development" is also highly contested, with some arguing that this ideology is in effect a Trojan horse for a neo-colonial or imperial project. For others, development can be both a project and a process. Despite the contested nature of the term, development has arguably become *the* central mission, rationale and justification for the existence of Third World states. Indeed, there has been a substantial amount written in recent decades on a purportedly new genus of states: developmental ones (Amsden 1989; Wade 1990).

For some scholars the nature and role of the state is the central one in international development. For example, Acemoglu and Robinson (2012) argue that where colonial institutions were more fully implanted, more successful development has been seen as a result of rule of law and property rights being more embedded, for example. This, they argue, happened in settler rather than non-settler colonies, which were more extractive. This perspective feeds into the debate in economics about whether institutions or geography matter more for developmental outcomes (Rodrik *et al.* 2004). However, this neglects that institutions are geography in the sense that they are both associated with geographical territories and what geographers call inter-scalar relations between the global and the local, for example, and help constitute them. More recently others have argued that it is policies, rather than institutions or physical geography, which matter most for economic development outcomes. However, this perspective might be questioned for its methodological nationalism and a-geographical theorisation. Relations between places shape policies, as do the particular socio-economic and physical geographies of nation-states. Nonetheless, the state remains a central "actor" or ensemble of institutions and practices in the (under)development process.

In the immediate post-colonial period for Africa and much of Asia there was extensive optimism about the role that the state could play in development. This was, in part, based on the successes of post-war rebuilding in Europe under the auspices of Keynesian welfare states. Some of the most influential theories about the role of the state in development included those of Rosenstein-Rodan (1943), Hirschman (1958) and Myrdal (1968). These different theories emphasised the importance of the state promoting development through planning and the expansion of manufacturing, in particular, which was seen to be more technologically dynamic, create higher added value and have greater employment potential. Rosenstein-Rodan's "big push" theory argued that both capital and consumer goods industries should be developed simultaneously to allow for complementarities or synergies, market creation and economies of scale. He argued that the creation of consumer goods industries would create a market for capital goods, and the workers employed in capital goods industries would create a market for consumer goods. More recently, Jeffrey Sachs also employs the idea of a "big push" in relation to increasing aid to avail of, or create, synergies between different sectors.

As also noted earlier, the Latin American structuralist school of economics argued that the imperative for economic development was to engage in import substitution of manufactured imports from the United States and Western Europe in order to diversify and develop economies. This assigned a central role to the state in the process of economic development and in many ways served as a precursor to later theories of the "developmental state". As noted earlier, ISI had problems unless complemented by measures to promote exports, as was the case in the successful NICs of Asia. In some of the most successful NICs stepwise or sequenced policies of primary import substitution of less technologically sophisticated products was followed by primary export promotion and then secondary import substitution of more technologically advanced and capital-intensive industries, and then secondary export promotion (Gereffi 1990).

Both modernisation theory and heterodox, or non-mainstream, structuralist approaches placed a heavy emphasis on the role of the state in development, and for geopolitical reasons – to prevent the advance of communism in the Global South – the World Bank adopted elements of these. For example, the World Bank was willing to selectively support policies of import substitution until the early 1980s, and under Robert McNamara that institution promoted state-led development strategies in the 1970s focussing on "basic needs", such as water provision and integrated rural development. Both modernisation theory and orthodox (or non-radical) structuralist approaches, which did not seek an overthrow of the capitalist system, shared an underlying assumption that states would be rational, Weberian actors which would promote development in the common interest. However, with sometimes poor performance and the generalised economic crises of the 1970s and 1980s, this optimistic view of the role of the state generally gave way to a more pessimistic or restricted view of the role that these institutions could or should play in promoting development.

Rethinking the state's role in development: "overdeveloped" and "neoliberal" views

After the initial post-independence optimism about the role which the state could play in guiding and promoting economic and social development, some social scientists began to question these earlier, positive, managerialist (Weberian) conceptions of the state. Neo-Marxian theories were developed which posited an alliance between domestic capitalist classes, bureaucratic elites and international capital which perpetuated economic underdevelopment, or in the case of Latin America "dependent development" (Evans 1979). These theories draw on Marx's analytical framework but argue that capitalism results in underdevelopment rather than development in the Global South. For Hamza Alavi (1972), the underdevelopment of the domestic bourgeoisie, or capitalist class, in post-colonial societies resulted in an "overdeveloped" state which was relatively autonomous in relation to its own society but was heavily influenced by the needs of, and responsive to, international capital, resulting in exploitative economic relations and underdevelopment. In relation to Africa, the "Kenyan debate" was about whether the state was primarily responsive to the domestic or international fraction of capital (Swainson 1980; Beckman 1980). Later work on Kenya emphasised the importance of the domestically based but "non-indigenous" bourgeoisie of Asian extraction (Himbara 1994).

While cynicism about the role of the state in the economy is often traced to the "Reagan and Thatcher revolutions" in the US and UK and the intellectual currents associated with these and their precursors, as noted earlier, there was also substantial left-wing criticism of actually existing states in the Third World. These concerns, from both left- and right-wing perspectives, were to come to a head with the Latin American debt crisis of the early 1980s.

As noted earlier, the global economic turbulence of the 1970s and the abandonment of the Bretton Woods system of fixed exchange rates, the second oil crisis and the Mexican debt default created conditions where many developing countries experienced economic crises and severe difficulty in accessing private international capital markets. Many countries were consequently forced to turn to the World Bank and the IMF for balance-of-payments support. In exchange the IFIs demanded SAPs be implemented. Fundamentally, at least in the initial stages, these were about reducing the role of the state in the economy so that it would function as a "night watchman" over the market, which it was felt would drive development.

The impacts of programmes of economic liberalisation varied depending on the specific geographical political economies undertaking them. They were generally more "successful", in their own terms, in countries in Southeast Asia, for example, where a regional product cycle was at work. Product cycle theory posits that industrial production moves from initial higher cost centres of innovation to lower-cost locations as technology diffuses and rents are dissipated. However, where this was not operative, they often resulted in competitive displacement of manufacturing production and social immiseration, particularly in much of Latin America and Africa during the lost decade(s) of the 1980s and part of the 1990s. "Infant industries" in these continents could often not compete with imports; cutbacks in health and education expenditures had devastating social and economic impacts, and a generalised "hollowing out" of the state associated with austerity resulted in dramatic reductions in development capacity. The devastating impacts of SAPs in these cases have been extensively analysed.

State failure, violence and underdevelopment

While the reasons for the general failure of SAPs in the developing world have been explored extensively in the literature (Lensink 1996; Mohan 2000), the effect on states varied depending on the economic context and the balance of social forces. In cases of previous state collapse, a "trough factor" (Martin 1993), where in effect the only choices were continued disorder or state reconstruction, was sometimes observed. In some cases "strong states" emerged, as in Ghana and Ethiopia, for example, which were able to take of advantage of inflows of hard currency associated with structural adjustment to relieve foreign exchange constraints on economic growth (Hutchful 2002). For example, if there are severe shortages of forex, spare parts for machinery cannot be imported and the economy grinds to a halt. In other cases, however, the cutbacks in government expenditure associated with fiscal rectitude or government spending cuts under SAPs resulted in the dissolution of the "patrimonial glue" which held some extant regimes together (Reno 1998). (Max Weber distinguished between rational-bureaucratic and patrimonial systems of authority. Patrimonialism is a kind of "who you know" politics, where state positions are awarded on the basis of personal contacts, rather than merit, for example.) In some instances these dynamics fed into the creation of trans-border crisis complexes in West and Central Africa, for example, which had their own enduring dynamics (Berg 2008).

From a rational choice perspective, which essentialises people as rational actors, Bates (2008) argues that state rulers in Africa are "specialists in violence" who may choose to deploy it either to prey on wealth or to protect its creation. Despite conditions of often intense poverty and the incentives which that creates for conflict, such as the lowering of opportunity costs to engage in violence for youth, African rulers tend to remain longer in power than those in other world regions. In part this is because of the way in which they are able to leverage external resources, such as aid or inflows of foreign currency from natural resources, into internal authority (Peiffer and Englebert 2012). In many developing countries regime maintenance, rather than economic and social development, is the primary goal of state regimes, despite public pronouncements to the contrary. This often results in unstable social formations, however, as state hegemony is not secure. The theory of hegemony was developed by the Italian Marxist, Antonio Gramsci. He sought to solve the puzzle as to why the working class did not revolt against their own exploitation, but rather actively participated in it. He proposed cultural-political hegemony as the reason. Hegemony is "coercion informed by consent" (Marais 1998), where capitalist social relations come to be seen as "common sense". This theory can also be applied to international relations where great powers attempt to present their interests as common interests in the international system through free trade, for example.

Developmental and catalytic states

The general failure of structural adjustment to achieve economic transformation in the developing world led to increased academic interest in a variety of other state models, particularly those of the NICs of East Asia. It was argued that these states had succeeded in industrialising by deliberately creating economic incentives which deviated from those of the "free" market – "getting prices wrong" (Amsden 1989), channelling finance to fast-growing industrial sectors and adopting a variety of other policy interventions to deliberately steer economic diversification. According to Amsden and Wade, developmental states are characterised by insulated, meritocratic yet embedded bureaucracies. Others characterise the social relations of developmental states as "embedded autonomy" (Evans 1995), where the state bureaucracy is linked into or responsive to but largely autonomous from, rather than captured by, societal interest groups, particularly private capital. This allows information to flow between the private sector and the state in addition to allowing state authority to be exercised to achieve developmental goals and aims. According to Woo-Cumings (1999, 27) "such economic and political relationships often imply a corporatist framework, involving large economic groupings (like the *keiretsu* [corporate conglomerates in Japan] or the *chaebol* [their equivalent in South Korea]) with which the state can coordinate and negotiate investment decisions".

In an evolution of this literature Linda Weiss (1998) argued that states in Europe, East Asia and elsewhere continue to play a vital role in the promotion of economic development. With the worldwide expansion of neoliberalism, she coined a new category of states – catalytic ones – which facilitate and guide private-sector development, rather than being highly directive as in the developmental state model. In light of the "revisionist thesis" promoted by authors such as Amsden and Wade, and partly at the instigation of the Japanese government, the World Bank also began to revise its conception of the appropriate role of the state in the economy in the 1990s, arguing that states should match "capabilities" with "roles" (Moore 1999). However, this ignored the possibility of how to build state capacities – one of the most important questions in international development. In practice, however, it has supported the work of the African Capacity Building Foundation, which works to develop national statistical or financial management capabilities, for example, rather than developmental states.

One of the key, and often neglected, factors in the emergence of developmental states is either an internal or external security threat which incentivises state elites to develop the economy as a source of legitimation and funding stream for the military. In Africa the developmental states of Rwanda and Ethiopia are, or were until recently in the latter case, dominated by minority ethnic groups who have experienced genocide, for example. However, this state strength may be implicated in the perpetuation and deepening of state weaknesses in other countries. For example, the Rwandan government has repeatedly invaded the neighbouring DRC and has been implicated in supporting rebel groups in the east of the country, although they deny this. This is another important reason to avoid the "territorial trap" (Agnew and Corbridge 1995) of methodological nationalism when considering state forms in the developing world – state strength and weakness are mutually imbricated and transnationally constitutive.

There are also a variety of models in development not captured by the state–market dichotomy. For example, some scholars have written about the role which civil society can play in the co-production of (public) goods and services (Tendler 1997). For Tendler, the key to effective, development-oriented governance may be "synergy" between different social forces and actors. She develops this argument through reference to the experience of the Brazilian state of Ceará.

Brazil has also been marked by state innovation at a national level through programmes such as "Zero Hunger" under Worker Party Rule from 2003 onwards. This programme had a variety of dimensions to it, including the creation of low-cost restaurants. During that time the Brazilian state could have been characterised as hybrid neoliberal/developmental or neostructuralist, where macro-economic policy followed economic orthodoxy, but microeconomic (firm level) and social policy were more interventionist. However, with the end of the global commodity super-cycle and recession in Brazil, Lula's successor to the presidency from the Workers' Party, and the first female to hold the post, Dilma Rousseff, was deposed. The new government which came to power was much more right wing under Michel Temer. His cabinet was composed entirely of men of European extraction, and his government passed a law freezing social spending in Brazil for 20 years. The presidential election of 2018 brought a former military officer, Jair Bolsonaro, to power. He has argued that the Chilean military dictatorship did not kill enough people and that the state should subsidise gun purchases for citizens. He has also argued that the poor should be sterilised to reduce crime rates.

Globalisation: the rise or retreat of the state in development?

The nature and role of the state in development is fundamentally related to globalisation. However, globalisation is not a static process, but a dynamic one involving rapid evolution in the forms of socio-spatial relations. The "rise of China" in the international political economy has reconfigured the nature of globalisation, given the still strong role of the state in the economy there and the prominent role which SOEs have played in that country's "go out" policy to promote overseas investment by its companies.

There has been a debate about the nature of the Chinese state. It has been variously characterised as developmental, neoliberal, Neo-Stalinist, market socialist or fragmented authoritarian (Lieberthal 1992). Even as it has liberalised its economy domestically, the Chinese government has expanded its spatial reach overseas through its state-owned corporations – respatialising rather than reducing the power of the state. This new model of globalisation, which is state promoted, not only calls into question the ontological distinction between "states" and "corporations" as separate social forces but also the posited opposition between the power of states and "globalisation". One of the most influential works which questioned the separation or distinction between state and society is Migdal (2001). Most of the 89 Chinese corporations in the Global Fortune 500 index of the biggest companies in the world are state-owned.

While many analysts now agree that states have a vital role to play in promoting economic and social development, the extent to which they can do this depends on their particular configuration. A central axis of debate is the nature of "the" state: whether it is hard (has the ability to enforce its will) or soft (ineffective and reluctant to challenge powerful social forces) (Myrdal 1968). Other axes of debate centre on whether or not states or societies are "strong" or "weak" in relation to each other. This approach also suffers from methodological nationalism by often not interrogating in relation to which social forces these designations are applied. For example, authoritarian states may be "strong" in relation to their domestic societies but weak relative to international capital.

Actor–network theory provides a more nuanced characterisation of states as assemblages of institutions, practices, actors and artefacts (things). Under conditions of globalisation, such a perspective allows us to examine the constitution of particular states and how they may be differentially and selectively internationalised (responsive to international forces), for example. Geographers have also recently theorised the spatiality of state networks and sovereignty using assemblage theory, arguing that under conditions of globalisation with a shift from "government" to "governance" by networks that a primary characteristic of states is that they possess "reach", not "height" (Allen and Cochrane 2010). In this theory states are not "national" but operate through multi-sited power relations and assemblages, and consequently development outcomes depend on the nature of these interactions and the power capabilities of different actors.

Different world regions have given rise to different theorisations of the state. For example, in relation to Africa much has been written about neopatrimonial, which combine patrimonial and Weberian rational – bureaucratic logics, or rhizome states which have patronage networks through society in a root-like fashion but fail to achieve state hegemony because of their inherently exclusionary nature (Bayart 1993). Other influential theories related to the African context include those of the "suspended state", which is largely disconnected from its own domestic society, giving it an outward orientation (Samatar and Samatar 1987), and the "bifurcated" state (Mamdani 1996) created under colonialism, where in broad-brush terms people in urban areas were governed as citizens, whereas in rural areas they were made into subjects with very constrained rights.

In the East Asian region the central theory in recent decades has been the developmental state. However, with the advent of network trade, where components for manufactured products are often sourced in multiple different countries and then assembled in a low-wage economy before being exported for consumption, some have questioned whether this model still applies in the original NICs (Yeung 2016). Others have also questioned the opposition between patrimonialism and developmentalism, arguing that some states in Africa are characterised by developmental patrimonialism (Kelsall 2013) – that is, that certain forms of corruption may concentrate rents, thereby allowing for capital accumulation and development.

Despite widespread neoliberalisation across much of the developing world, the role of the state in promoting economic diversification and improved living standards and the ways in which it may hamper or prevent these outcomes are still some of the most debated topics in the international development literature. A recent book argues that the development divergence between Asia and Africa can be explained in voluntaristic terms – that Asian leaders wanted to promote development through the upliftment of the majority of the rural population, whereas this was not the case in Africa (Henley 2015). However, this neglects both what are called "selection environments" and "political opportunity structures".

Selection environments refers to the socio-economic conditions which give rise to certain social or economic characteristics to succeed over others. For example, Russia as a sprawling multi-national country (some would say empire), and given its distinctive history, has tended to have authoritarian leaders. "Russia is the biggest country in the world, twice the size of the USA or China, five times the size of India, seventy times the size of the UK" (Marshall 2015, 9). We can think of this as the selection environment, a concept which was initially developed in the biological sciences to explain why certain plant species thrived in certain places and not others. Political opportunity structure refers to the room for action that political actors have or face given prevailing distributions of power, which will lead them to follow certain courses of action over others. This feeds into the debate about the relative importance of structure and so-called agency in determining social or developmental outcomes. Do people or politicians really have "free will" to choose different courses of action?

In psychology, some theorists have cast doubt on the mind/brain division that much of international relations theory implicitly accepts (Green 2003). For those theorists, the brain is simply another bodily organ which responds to physical and chemical stimuli and has learnt from past experience, and there is no separate "mind". This is not to dispute that some people have power over others, but rather to question the extent to which free will or voluntarism can be attributed to actions. In this context, it then becomes very important to distinguish between agency and power. For example, although some African political elites may have recently gained power in relation to "external" actors, their power should not be confused with their agency or lack thereof. While certain politicians, such as Vladamir Lenin in Russia, believed that societies could be remade through acts of will, this conclusion was an outcome of all of the events and thoughts leading up to it. As the noted political scientist, Alexander Wendt, reminds us, social structures and other things like commodities originate as ideas, but these ideas are not autonomous, instead arising from particular upbringings, contexts and biologies. States are, however, reconfiguring discourses and global imaginaries.

The "rise of China" and "the South" more generally is creating a new macro-geopolitical region – a "South Space" – which has greater policy latitude and autonomy from the IFIs in some cases (Carmody 2013a). This greater policy autonomy which countries across much of the Global South are experiencing may not fundamentally challenge neoliberalism, but rather lead to its renegotiation through resource nationalism, whereby resource-rich states capture greater shares of resource rents and revenues. Some refer to this conjunction of events as providing scope for neodevelopmentalism.

Geopolitics and developmental states

There are a variety of definitions of what constitutes geopolitics and geoeconomics (Lee *et al.* 2017). One way to think of geopolitics is that it is how power differences get expressed in the international arena spatially. How do certain social actors and countries/states dominate others economically, politically and militarily in the international political economy? What explains this uneven distribution and replication of this spatial structure of power? This has important implications for development, as will be explored next.

Without doubt, the most successful region of the world in terms of development since the 1960s has been East Asia. In contrast to the predictions of dependency theory, many countries in the region have managed to rapidly industrialise and lift most of their populations out of poverty. A huge volume of literature has been written to try to explain how this has happened. As noted earlier, one recent book compares the developmental trajectories of Asia and Africa and largely ascribes the difference to the intentions of political elites (Henley 2015). This does not explain why elites were supposedly developmental in Asia but not in Africa, however, and also neglects the importance of state structures, state–society interactions, geopolitics and other factors.

Many analyses attribute the rise of East Asia to the presence of so-called developmental states in the region. According to one observer:

> Fundamental to East Asian development has been the focus on industrialization as opposed to considerations involving maximizing profitability on the basis of current comparative advantage. In other words, market rationality has been constrained by the priorities of industrialization. Key to rapid industrialization is a strong and autonomous state, providing directional thrust to the operation of the market mechanism. The market is guided by a conception of long-term national rationality of investment formulated by government officials. It is the "synergy" between the state and the market which provides the basis for outstanding development experience.
>
> (Öniş 1991, 112)

In many ways Japan provided the model for this type of approach (Johnson 1982), as it was never colonised, unlike most of the non-Western world, and it industrialised in the nineteenth century after throwing off the strictures of Western-imposed free trade, which had been enforced by so-called "gun-boat diplomacy". In 1853 Commodore Perry of the US Navy led a small fleet of ships into Edo Bay, now called Tokyo Bay, in order to end Japanese self-imposed isolation and open the country up to trade with the US, by force if necessary (Stavrianos 1981). In order to make the point Perry fired volleys of blank cannon shot, supposedly to celebrate American independence, and threatened to vanquish the Japanese militarily. The Meiji Restoration of 1868, however, gradually restored autonomy to Japan and enabled it to industrialise by importing Western technology, methods and technologists. On coming to power the new emperor said that "knowledge shall be sought from all over the world and thereby the foundations of the imperial rule shall be strengthened" (quoted in Kissinger 2011, 79).

While the American state has promoted economic liberalisation around the world, at least since the 1980s, it continues to operate what Mariana Mazzucato (2011) calls a "hidden developmental state" itself through subsidies for technology development in the arms industry, for example, which can then be further developed for civilian application. Recent work has also shown how the American occupation of Japan after World War II, and its strictly planned nature, laid the groundwork for its subsequent economic resurgence (Glassman 2018). American involvement was also important in the South Korean case subsequently (Glassman and Choi 2014), as it supplied most of its foreign exchange through aid at certain points in time.

Neoclassical economists argue that the role of the state should be as limited as possible in economic development and that the "free" market, left to its own devices, will find the most efficient allocation of resources based on relative scarcity as expressed through prices. However, Wade (1990) characterised the Taiwanese experience as one of "governing the market" – that is, where the state does not supplant the market as an economic coordination mechanism but strategically manages it in order to promote industrialisation.

According to Leftwich (2000) a developmental state is one which has development as its overriding priority and subordinates other policy goals or objectives to that. However, in general, to speak of developmental states is a misnomer. Most governments and regimes are concerned above all else with state or party maintenance or national security and survival. Jessop (2002) notes that security may override capital accumulation as a state priority. This can be seen, for example, in the extensive state planning which was implemented in Britain during the Second World War. The most successful "developmental states" in Asia and Africa have not necessarily had development as their overriding priority, but have in reality been security states, pre-occupied with regime and/or state survival. This applies to Taiwan, South Korea, China, Ethiopia or Rwanda, although Himbara (2016), a former government advisor there, has questioned the latter country's status as a developmental state, suggesting many of the statistics which would support this interpretation have been deliberately falsified, although this finding has been strongly disputed and debated (see Ring 2018; Anonymous 2018). These have also been authoritarian states, capable of disciplining or repressing labour and the peasantry, but also what is typically amongst the most powerful social forces or actors in given state–society formations – (trans)national capital.

So-called developmental states typically engage in what neoclassical economists call "repressing" finance or regulating it to channel investment to productive economic activities. However, if a security threat is not imminent, often there are easier, more lucrative or otherwise more contextually incentive compatible ways of governing. The implications of this are stark. While some have argued that systemic vulnerability promotes economic up-grading (Doner *et al.* 2005) or that it may be useful to manufacture security threats to have states behave in more developmental ways, this seems unlikely to occur. Why would foreign powers do this? Furthermore, the extent to which "the state" is inter-subjectively understood to be an entity worth defending and strengthening varies from context to context (Dunn 2001). Where the state has been experienced as an agent of repression, people may choose the exit option and disengage from it by reverting to subsistence agriculture, as in parts of Africa during the 1980s, for example (Cheru 1989).

In terms of the structure of developmental states, both Amsden and Wade emphasised that they were relatively autonomous from domestic social forces and that this autonomy enabled them to set appropriate economic policies using targets and incentives, such as export competitions, and sanctions. In many other countries of the Global South states are often captured by powerful domestic social interests (Migdal 1988), which may compromise their ability to devise economic policies which are broadly socially beneficial. As noted earlier, Peter Evans (1995) extended Amsden's and Wade's insights through the development of the concept of "embedded autonomy". This was the idea that effective states are sufficiently autonomous from business or capital to be able to discipline it by insisting on good export performance, for example, in return for subsidies. However, in order to be responsive to the needs of business in terms of training, penetrating export markets, etc., the state also had to be embedded with business but not captured by it.

The theory of the developmental state has also come under critique. Chang (2013, 85), for example, argues that "the predilection to set up an opposition between state and market has resulted in 'downplaying the role of class' in analysing development" and that

> the concept of the developmental state can be derived only with a particular understanding of labour that is disempowered and depoliticised. Statist development policies are essentially anti-labour in that they 'fetishise' or mystify the state as if it exists apart from social relations in 'facilitating' development.

Rather he argues the developmental states of East Asia were capitalist:

> Emerging discontent from labour and its disruptive potential, alone or in combination with other forms of social unrest was a key factor in East Asia that prompted individual capitals at an early stage to concede to state coordination in exchange for its tight control over collective labour. . . . Indeed, the strength and characteristics of the East Asian states, and the illusion of their autonomy, are a consequence of the isolation and suppression of workers and social movements.
>
> (Chang 2013, 85)

What makes a state developmental?

As noted earlier, there have been a variety of different theories about what makes a state developmental. Core components seem to be that the state bureaucracy be meritocratic and competent and that policy makers be insulated from the demands of special-interest groups. However, the political opportunity structure is also key. For example, the confusion and dislocation created by war may provide state elites with an opportunity to reshape society, as was the case in South Korea. However, external factors may also be very important. For example, Pempel (1999) argues that late industrialisers in East Asia benefitted from a "developmental regime". This was where the United States allowed largely free access to its market and provided very substantial ODA for Taiwan and South Korea to ward off the threat of communist takeover in those countries. However, many countries in Africa received substantial development assistance from the US, such as Zaire under the notorious dictator and kleptocrat, Mobutu Seso Seko, which did not industrialise.[1] Thus, it is the interrelationship between internal and external factors which is key. According to Robert Wade at the London School of Economics development is more like a combination lock than a padlock.

In order to stay in power, regimes in the Global South may offer more powerful actors in the international political economy an "extraversion portfolio" from which they can benefit (Peiffer and Englebert 2012). Extaversion implies an external orientation, either in politics or economics. In economics an extraverted economy meets the needs of people overseas rather than domestic citizens' needs. For example, the Ethiopian People's Revolutionary Democratic Front (EPRDF) is known to be a very repressive government and yet it continues to attract very substantial amounts of Western ODA (Fantini and Puddu 2016). In part this is because the country is seen to be successful at poverty reduction, with the economy quadrupling in size between 2005 and 2015 (Singh and Ovadia 2018) and because the regime has brought relative stability to the country after a long-running civil war. The Western powers have supported Ethiopia because of its geopolitical importance. According to a British military official in Ethiopia, the government there is also willing to fight "proxy wars" on the West's behalf by invading Somalia to depose the Islamic Courts Union government in 2006, for example (in conversation, Addis Ababa 2011).

Developmental states in Africa?

Oftentimes social scientists or those in the media talk about Africa as if it is one place, using phrases such as "Africa is poor". We can think of this as a form of methodological continentalism, which disguises the huge variations in socio-economic conditions across the continent, even if it is true as a general statement. Mauritius, for example, has a highly developed and diversified economy, and Sandton in Johannesburg is extremely wealthy, as are parts of Accra in Ghana (Figures 5.1 and 5.2).

Increasing numbers of people live in poverty in Africa (Beegle *et al.* 2016). This is, in part, a result of the continent's continuing deindustrialisation (UNCTAD 2012a). However, there is a largely unexplored paradox in African development: there are many examples of successful manufacturing and service firms and some successful industrial policies on the continent. The most successful industrial policy in recent years on the continent has been that of Ethiopia, which will be explored in more detail later.

FIGURE 5.1 Norton Rose Fulbright South Africa building, Sandton, Johannesburg

FIGURE 5.2 High-end housing in Accra, Ghana
Photo credit: Ian Yeboah

While Africa is often associated with developmental failure, how do successful, internationally competitive firms and foreign investments emerge in Africa, and what lessons can be learnt from their experiences? This is an important time to examine this question, given the end of the commodity super-cycle and generally improved governance and business environments on the continent (Rotberg 2013). These, combined with dramatic changes in the structure of global trade relations, particularly the increased importance of GVCs and associated "network" trade in tasks, such as assembly, rather than in complete products (Kaplinksy and Morris 2016), offer potential for African industrial development. Countries such as China, Malaysia and Thailand are increasingly concentrating on medium- and high-tech manufactured exports (Newman *et al.* 2016), potentially opening up product space for Africa to compete in lower-end manufacturing. By way of example, 13 per cent of apparel chief procurement officers interviewed recently ranked Ethiopia as a top-three sourcing destination over the next five years (Berg *et al.* 2015). Philip Van Huesen, the owner of the Tommy Hilfiger and Calvin Klein brands, has begun producing clothing at scale in a major new industrial park in Hawassa, a few hundred kilometres outside of the Ethiopian capital of Addis Ababa, which is meant to employ 60,000 people. However, if endogenous, or internally propelled, industrialisation is to take hold, African companies need to meet Asian competition, create ties to international markets with high-import propensities, such as China, and overcome other challenges to growth and innovation, such as infrastructure and access to finance.

The central problematic of Africa's development is rooted in its economic structure, with not a single (Sub-Saharan) African country having a manufactured product in its top three exports, according to the United Nations' Comtrade data (Cramer 2016), excluding Mauritius. For some, the recent commodities boom seemed to preclude the need for more active industrial strategies on the continent (Thakkar 2015). However, despite rhetoric about the continent "rising" or becoming a "global powerhouse" (Bright and Hruby 2015), a large part of Africa's continuing developmental impasse is the general absence of territorial innovation systems, with some exceptions such as the new media cluster of firms in Cape Town (Booyens *et al.* 2013). Territorial innovation systems combine growing firms and training institutions to enable new higher-value products and processes to be developed. However, the continent as a whole remains dependent on primary commodity exports.

Primary commodity production tends to be of relatively low productivity and generates fewer technological breakthroughs, linkages and consequently multipliers than manufacturing and some services. Furthermore, primary commodities are notoriously volatile in terms of their prices, and the recent super-cycle or boom notwithstanding, the long-term trend since the 1930s for primary commodity prices continues downwards (Moseley 2014), with the exception of oil. This trend gives added impetus to the imperative to move beyond dependence on them.

The development policy challenge of industrialisation

Given the intense competition in the global economy between industrial regions and pervasive economies of scale, which often make bigger plants more efficient and positive externalities, where knowledge spills over between co-located firms, for example, and other handicaps such as the underfunding of universities in Africa, the establishment of territorial innovation systems has proven very difficult there. However, while Africa as a whole performs poorly on traditional measures of innovation, such as patents or numbers of scientific and technical publications (Oluwatobi *et al.* 2015), there are nonetheless successful manufacturing and service firms on the continent.

Quantitative research has shown a relationship between production, innovation and growth amongst small firms on the continent (cf. Goedhuys and Sleuwaegen 2010). However, such approaches tend to treat firms as ontologically separate units of analysis, thereby often ignoring how networks and social relations may have contributed to firm growth and development. Rather than conceptualising firms in this way, they can also be conceived of as nodes within networks, such as those of global production, trade or national political economy (Ouma 2015). Territorially embedded development processes can be thought of as outcomes of transnational and transnationalising assemblages of artefacts and actors who have different skills, capabilities, subjectivities and interests which are shaped by their shifting positionalities in networks. Consequently, examining the role of networks is vitally important. How and why do such successful firm networks emerge, and what policy lessons can be drawn from their experience more broadly for Africa and beyond?

The positive relationship between industrialisation and development is well established; however, there are a variety of definitions of what constitutes industrialisation in the literature. For the purposes of this book, it entails the sustained raising of productivity across sectors and the capturing of this value locally and nationally. This is often an inherently difficult process given the established competitive advantages, such as capabilities and economies of scale, already developed by regional production complexes in other parts of the world.

Increases in manufacturing output or productivity are not by themselves enough, however. While manufacturing output in the Global South now exceeds that in the Global North, average incomes in the South remained roughly stagnant as the proportion of those in the North, at around 5 per cent (Arrighi *et al.* 2003 cited in Selwyn 2014). Some of the most efficient or productive transnational corporation plants in the world are in the Global South, such as Ford car plants in Mexico. However, wages remain low and profits flow primarily back to foreign shareholders.

In Mexico foreign-owned factories which get tax breaks on imports that are used to assemble products for exports are known as *maquiladoras* and primarily employ young women, often under extremely exploitative labour conditions (Cravey 1998). However, low wages and often more lax environmental regulation are attractive to transnational corporations, though they are not necessarily positive from a developmental point of view. "Between 1980 and 1997 Mexico's share in world manufactured exports rose tenfold, while its share in world manufacturing value added fell by more than one third and its share in world income (in current dollars) [fell] by about 13 percent" (UNCTAD 2002, 80 quoted in Hart-Landsberg 2013, 85–86). In 2010 inflows of foreign direct investment to "developing" countries were more than half of the global total for the first time (UNCTAD 2012b), although there are also other attractions than cheap labour, such as land and natural resources.

In recent years there has been a substantial resurgence of interest in industrial policy as a mechanism through which to achieve "late development" (Stiglitz *et al.* 2013; Noman and Stiglitz 2015; Clark *et al.* 2016). A number of important theoretical innovations have also been made in recent years in understanding the nature of multi-scalar development in the context of deepening globalisation, such as product space (Hidalgo *et al.* 2007), political settlements (Khan 2013) and global production network (GPN) theories (Coe and Yeung 2015). In particular, GPN theories have focussed on the importance of "strategic couplings" between regional assets, such as skilled labour, and so-called lead firms (TNCs) (Yeung 2016). These strategic couplings involve win-win outcomes for the actors involved through capability development, for example.

Africa has also seen an increasing growth of indigenous TNCs in recent years. While it is well known that many major corporations have originated in South Africa, recent research has revealed that there are now a multitude of other African originating companies operating transnationally on the continent (Rolfe *et al.* 2015). Some prominent examples of these include the Nigerian-based Dangote group, which has operations across the continent and grew initially on the basis of extensive protection in its home market for cement, and the Madhvani group of Uganda, which is involved in sugar and chemical production amongst other subsectors. This also includes Africa's only "unicorn" technology company (a start-up valued at more than US$1 billion) – the Africa

Internet Group from Nigeria. However, a somewhat disturbing trend is that "African MNCs [multi-national companies] have targeted their FDI [foreign direct investment] activities mainly at other African countries, including the conflict-affected and fragile states" (Ibeh 2015, 135), creating regional rather than GPNs. These investments appear to be mostly market serving rather than efficiency seeking – a form of economic introversion also visible in other sectors on the continent, such as business process outsourcing (BPO) (Mann and Graham 2016). This inherently limits growth potential by militating against a "leap into global network trade", which is potentially important to the continent's economic development. However, there are also some counter-examples of Mauritian clothing companies manufacturing in Madagascar for global buyers (Morris *et al.* 2015) and of smaller companies, such as Sole Rebels in Ethiopia, which now sells its shoes in 30 countries around the world (Nsehe 2012).

The "inclusionary bias" of GPN theory, or the tendency to focus research on areas which are plugged into them, has recently been noted (Bair and Werner 2011). However, theoretically there is a need to explain alternative modes of globalisation of firms and regions, such as direct exporting to end consumers, currently largely excluded from the gaze of GPN research. Often small and medium-sized enterprises from Africa export directly to end consumers or through intermediaries, rather than being embedded in GPNs with lead firms. For example, 40 per cent of Kenyan manufacturers export (Newman n.d.), although these are often to regional markets. Small and medium-sized enterprises generate the most jobs, globally and in Africa, and there is a need to focus on more "ordinary" firms, rather than just high-tech ones or those connected into GPNs.

How do successful learning and exporting firms and investment-development pathways emerge and stabilise in Africa? Sometimes these are based on "frugal" innovations and involve different forms of connection to the global economy than that posited or examined in the GPN literature. Examples of frugal innovation include using smart phones and cameras to manage production lines remotely in a furniture factory in South Africa (Carmody 2014) to designing and building an aircraft from scratch using locally available materials (Muchie *et al.* 2003).

A cluster is generally defined as a geographic agglomeration of firms producing the same types or products, perhaps in combination with those in related industries. Theoretically, while the specifics behind each firm and cluster experience vary, successful industrial development experiences in Africa are often a result of multiple strategic couplings, including those between firms and, sometimes the state, described in more detail later, combined with locally innovative firm practices and technological adaptations, otherwise known as technology diffusion management. Such productive couplings or "institutional thickness" (Amin and Thrift 1994) have been largely absent in Africa, or in some cases the couplings between state and firms has been unproductive or obstructive (Hampwaye and Jeppesen 2014; Urassa 2014).

Industrial systems as assemblage and the role of the state

What kinds of social relations promote industrial development? While there has been substantial academic work done on the role of the state in the promotion of successful industrial development and its role remains central, new types of social relations and assemblages are emerging and emergent which may promote industrialisation. Ethiopian manufacturing has been growing at an exceptionally high rate; 15 per cent a year in five recent years (Tomkinson 2016), and the reasons behind this will be discussed in more detail later.

Multiple and effective strategic couplings can be conceived of as an assemblage (see Figure 5.3). For example, one of the most recent successful industrial development experiences on the continent – the Huijan shoe company investment in the Bole Lemi special economic zone in Ethiopia (Staritz *et al.* 2016) – involves foreign investment and is an outcome of a transnational assemblage involving multi-axis strategic couplings between foreign investors, regional assets, government and a "third-party state" – China. As Brautigam *et al.* (2015) note, "Ethiopia has adopted an active, state driven industrial policy aimed at incentivising exports, attracting lead firms and foreign direct investment (FDI), supporting local firms, and creating local linkages to promote priority sectors such as apparel and textiles". Eighty per cent of leather used by Huijan in Ethiopia is locally sourced (Hauge 2016), and the company recently announced

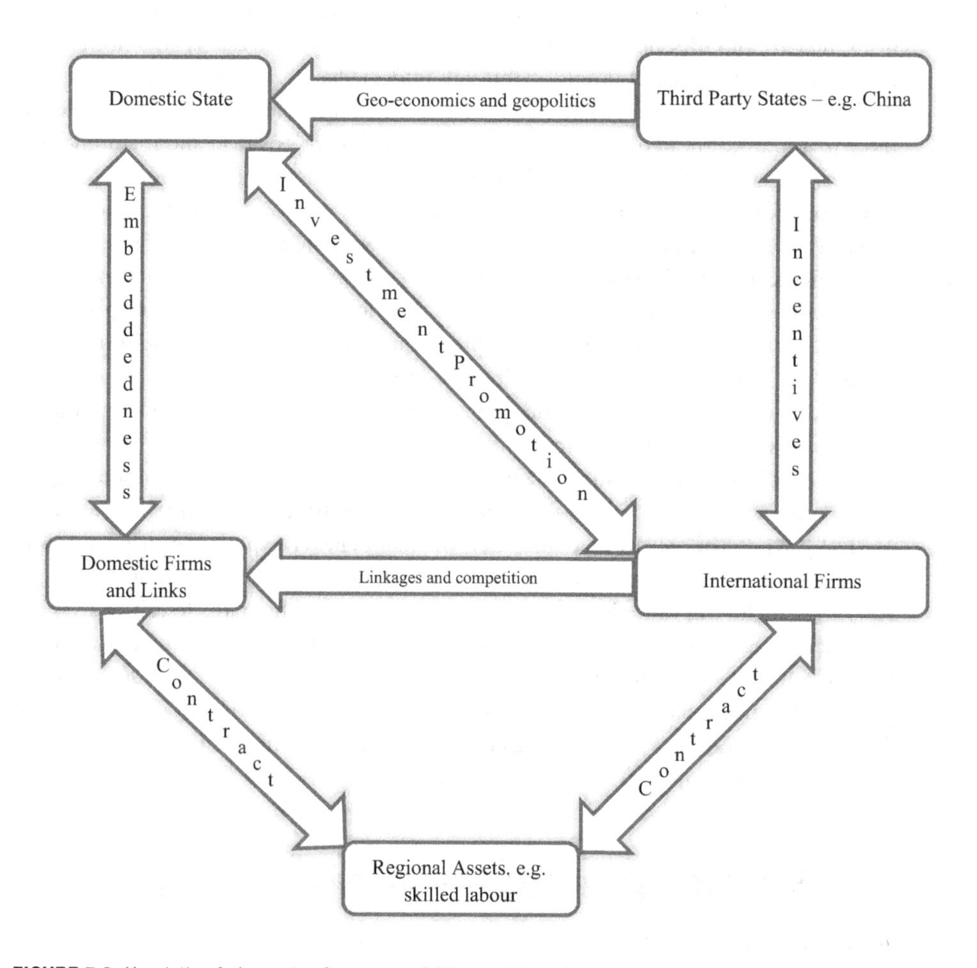

FIGURE 5.3 Heuristic of elements of a successful industrial system

that it will be moving its headquarters there from China, although some have reservations about the broader developmental potential of the Chinese-built special economic zones (SEZs) in the country. For example, Giannecchini and Taylor (2018) argue that because the Eastern Industrial Zone outside of Addis is completely occupied by Chinese companies, there is unlikely to be any technology transfer to local companies.

The "endogeneity factor", where better-performing economies attract more foreign investment, may also be operative in Ethiopia (Moghalu 2014). It not yet a member of the WTO (a form of deliberate non-assemblage), which allows it to engage in strategic economic interventions to promote industrialisation. Productive assemblages are, in turn, composed of sub-assemblages of actors and artefacts within firms, for example; however, the role of the state remains central.

Africa has the highest inter-firm productivity differentials of any world region (Newman *et al.* 2016), and this may partly reflect connections between large businesses and political elites, which may create certain advantages for some of these firms (Tvedten *et al.* 2014). State–firm relations and networks are then important, rather than simply assuming the state to be a barrier to enterprise development. In terms of developmental impact, what is important is the scale, productivity growth, rent (above market rate of return) capture and local and national linkages and untraded inter-dependencies created, such as knowledge spill-overs as workers and managers move between firms (Storper 1997).

To date, explanations for the relative success or failure of industry and technology policies in Africa have concentrated on the impacts of economic liberalisation (Stein 1992) and/or political settlements (the balance of power between political elites, bureaucrats and private-sector interests and the institutional arrangements put in place to reflect and stabilise these) (Whitfield *et al.* 2015). Political settlements are structured by the nature of economies and how these interact with past histories of conflict, identity and other contextual factors. They are also influenced by forces of "globalisation", often originating from outside the national territory. However, the political settlements literature still has an incomplete methodologically nationalist focus and consequently needs to be complemented and expanded through its deployment and engagement with other theories, such as product cycle theories.

The most poorly capacitated states, at least in terms of directing economic transformation, are often found in the poorest countries, where regime maintenance is often the overriding government priority. State capacities can be developed (UNECA 2016), and developmental ambition may be an important spur to this (Cramer 2016), as the relatively successful case of Ethiopia has shown (Chandra 2013; Oqubay 2015). Although current Ethiopian development is also associated with accumulation by dispossession. The global commodification of food and related price increases have spurred land grabbing in the country, creating labour, which is an enactment of global commodification chains. Land rights have recently sparked protests in the country, a government crackdown and then a change in prime minister. If conflict were to substantially escalate, it might undermine industrial development. Indeed the new prime minister, Abiy Ahmed, seemed to be close to being overthrown by protesting soldiers in 2018 (his getting them to do push-ups with him seemed to deflect them). "Developmental states" are often highly authoritarian. For example, the South Korean state massacred workers in Gwangju in 1980, which is often forgotten in Western countries.

When development is initially successful, this may turn into a recursive feedback loop as the realisation of profit through exchange feeds directly back to state power and capabilities development. Profits from production augment state power through the taxes levied on labour and companies during the production and circulation processes and through processes of learning-by-doing industrial policy. Thus, what is often seen to be a form of globalisation disconnected from the state – namely, the global sale of commodities – in fact represents a fused form of "commodity power" which may benefit not only corporations but also the states in which they are headquartered or where production takes place. Whether states prioritise development, however, depends on the nature of their social bases and issues such as whether or not they face a security threat which encourages prioritisation of national economic development.

The political opportunity structure facing states plays a critical role in determining whether or not they adopt a "nurture capitalism" or anti-developmental, neopatrimonial, regime maintenance stance (Van de Walle 2001). As noted earlier, according to the World Bank African states need to match capabilities and roles – providing basic infrastructure and macro-economic stability, for example, while otherwise allowing the "free" operation of the market. However, as some have noted, markets must be created before they can be freed.

Typically in World Bank studies the African state is viewed as a constraint on enterprise development to be removed, rather than conceptualised as an institution which can actively guide and promote economic transformation. However, as described earlier, it is the nature of the state, the political settlement and the types of assemblages which are constructed and the way in which the political economy is articulated to global regimes or not, such as that of the WTO, which determine whether or not industrial policy will be successful in terms of raising peoples' standards of living generally. However, others argue that "from the twin crises of global urbanization and industrialization signified by 'surplus humanity', to the faltering productivist behemoth of industrial agriculture . . . there are good reasons for considering that an epochal crisis may well be on the horizon" (Moore 2015, 27).

Most people around the world might think that industrialisation and urbanisation are good things. However, industrialisation is resource intensive and often, in the initial stages at least, highly exploitative of labour. Furthermore, the industrialisation of some world regions puts competitive pressure on others, sometimes resulting in their deindustrialisation – creating "surplus people" who do not function as effective producers for, or consumers in, the globalised economy. Is the "development" enterprise then fundamentally flawed? The next chapter explores a number of theories which suggests that it is.

Note

1 His given name was previously Joseph Mobutu, but on ascending to the presidency he assumed his new name which had at least two translations. "The Ngbandi translation reads, 'the warrior who knows no defeat because of his endurance and inflexible will and is all powerful, leaving fire in his wake as he goes from conquest to conquest'. In Tshiluba, the name translates to 'invincible warrior, Cock who leaves no chick intact'" (Callaghy 1979, 341 quoted in Dunn 2001, 239). Mobutu ran the country for over 30 years, until he was eventually deposed by a rebellion.

Further reading

Books

Amsden, A. 1989. *Asia's Next Giant: South Korea and Late Industrialization.* Oxford and New York: Oxford University Press.

Wade, R. 1990. *Governing the Market: Economic Theory and the Role of the State in East Asian Industrialization.* Princeton, NJ: Princeton University Press.

Articles

Öniş, Z. 1991. "The Logic of the Developmental State." *Comparative Politics* 24 (1): 109–26.

Singh, J.N., and J.S. Ovadia. 2018. "The Theory and Practice of Building Developmental States in the Global South." *Third World Quarterly* 39 (6): 1033–55.

Websites

Heterodox Economics newsletter, www.heterodoxnews.com/HEN/home.html

The United Nations Economic Commission for Africa, www.uneca.org

The United Nations Economic Commission for Latin American and the Caribbean, www.cepal.org/en

References

Acemoglu, D., and J. Robinson. 2012. *Why Nations Fail: The Origins of Power, Prosperity and Poverty.* London: Profile.

Agnew, J., and S. Corbridge. 1995. *Mastering Space: Hegemony, Territory and International Political Economy.* London: Routledge.

Alavi, H. 1972. "The State in Post-Colonial Societies: Pakistan and Bangladesh." *New Left Review* 74: 59–81.

Allen, J., and A. Cochrane. 2010. "Assemblages of State Power: Topological Shifts in the Organization of Government and Politics." *Antipode* 42 (5): 1071–89.

Amin, A., and N.J. Thrift, (eds.) 1994. *Globalization, Institutions, and Regional Development in Europe.* Oxford: Oxford University Press.

Amsden, A. 1989. *Asia's Next Giant: South Korea and Late Industrialization.* New York and Oxford: Oxford University Press.

Anderson, B. 1983. *Imagined Communities: Reflections on the Origin and Spread of Nationalism.* London: Verso.

Anonymous. 2018. "Rwanda's House of Sand: Brutality, Lies and Complicity." http://roape.net/2018/07/26/rwandas-house-of-sand-brutality-lies-and-complicity/.

Arrighi, G., B.J. Silver, and B.D. Brewer. 2003. "Industrial Convergence, Globalization, and The Persistence of the North-South Divide." *Studies in Comparative International Development* 38 (1): 3–31.

Bair, J., and M. Werner. 2011. "The Place of Disarticulations: Global Commodity Production in La Laguna, Mexico." *Environment and Planning A* 43 (5): 998–1015.

Bates, R.H. 2008. *When Things Fell Apart: State Failure in Late-Century Africa.* Cambridge: Cambridge University Press.

Bayart, J.-F. 1993. *The State in Africa: The Politics of the Belly.* London: Longman.

Beckman, B. 1980. "Imperialism and Capitalist Transformation: Critique of a Kenyan Debate." *Review of African Political Economy* 7 (19): 48–62.

Beegle, K., L. Christiaensen, A. Dabalen, and I. Gaddis. 2016. *Poverty in a Rising Africa.* Washington, DC: World Bank.

Berg, P. 2008. "A Crisis-Complex, not Complex Crises: Conflict Dynamics in the Sudan, Chad and the Central African Tri-Border Area." *Internationale Politik und Gesellschaft* 4: 72–86.

Berg, A., S. Hedrich, and B. Russo. 2015. "East Africa: The Next Hub for Apparel Sourcing?" www.mckinsey.com/industries/retail/our-insights/east-africa-the-next-hub-for-apparel-sourcing: McKinsey and Company.

Booyens, I., N. Molotja, and M. Phiri. 2013. "Innovation in High-Technology SMMEs: The Case of the New Media Sector in Cape Town." *Urban Forum* 24 (2): 289–306.

Brautigam, D., T. Weis, and T. Xiaoyang. 2015. *Ethiopia's Industrial Policy: The Case of the Leather Sector.* Mimeo.

Bright, J., and A. Hruby. 2015. *The Next Africa: An Emerging Continent Becomes a Global Powerhouse.* New York: St. Martin's Press.

Carmody, P. 2013. *The Rise of the BRICS in Africa: The Geopolitics of South-South Relations.* London: Zed Books.

———. 2014. "The World Is Bumpy? Power, Uneven Development and the Impact of New ICTs on South African Manufacturing." *Human Geography* 7 (1): 1–16.

Chandra, V. 2013. "How Ethiopia Can Foster a Light Manufacturing Sector." In *The Industrial Policy Revolution II: Africa in the 21st Century*, eds. J. Stiglitz, J. Lin and E. Patel. Basingstoke and New York: Palgrave Macmillan.

Chang, D. 2013. "Labour and the 'developmental state': A Critique of the Developmental State Theory of Labour." In *Beyond the Developmental State: Industrial Policy in the Twenty-First Century*, eds. B. Fine, J. Saraswati and D. Tavasci. London: Zed Books.

Cheru, F. 1989. *The Silent Revolution in Africa: Debt, Development and Democracy.* London: Zed Books.

Clark, D., L. Lima, and C. Sawyer. 2016. "Stages of Diversification in Africa." *Economics Letters* 144: 68–70.

Coe, N., and H. Yeung. 2015. *Global Production Networks: Theorizing Economic Development in an Interconnected World.* Oxford and New York: Oxford University Press.

Cramer, C. 2016. "Guns and Roses: Crossing the River by Feeling for Stones in Ethiopian Industrialisation – At Speed." In *Africa's Turn to Industrialize? Shifting Global Value Chains, Industrial Policy and African Development.* London: School of Economics and Political Science.

Cravey, A. 1998. *Women and Work in Mexico's Maquiladoras.* Lanham, MD and Oxford: Rowman & Littlefield.

Doner, R.F., B.K. Ritchie, and D. Slater. 2005. "Systemic Vulnerability and the Origins of Developmental States: Northeast and Southeast Asia in Comparative Perspective." *International Organization* 59 (2): 327–61.

Evans, P. 1979. *Dependent Development: The Alliance of Multinational, State, and Local Capital in Brazil.* Princeton, NJ and Guildford: Princeton University Press.

———. 1995. *Embedded Autonomy: States and Industrial Transformation.* Princeton, NJ and Chichester: Princeton University Press.

Fantini, E., and L. Puddu. 2016. "Ethiopia and International Aid: Development Between High Modernism and Exceptional Measures." In *Aid and Authoritarianism in Africa: Development Without Democracy*, eds. T. Hagman and F. Reyntjens. London: Zed Books, pp. 91–118.

Friedmann, T. 1999. "A Manifesto for the Fast World." www.nytimes.com/books/99/04/25/reviews/friedman-mag.html.

Gereffi, G. 1990. "Paths of Industrialisation: An Overview." In *Manufacturing Miracles: Paths of Industrialization in Latin America and East Asia*, eds. G. Gereffi and D.L. Wyman. Princeton, NJ: Princeton University Press.

Giannecchini, P., and I. Taylor. 2018. "The Eastern Industrial Zone in Ethiopia: Catalyst for Development." *Geoforum* 88: 28–35.

Glassman, J. 2018. *Drums of War, Drums of Development: The Formation of a Pacific Ruling Class and Industrial Transformation in East and Southeast Asia, 1945–1980.* Leiden and Chicago, IL: Brill Press and Haymarket.

———., and Y. Choi. 2014. "The Chaebol and the US Military-Industrial Complex: Cold War Geo-Political Economy and South Korean Industrialization." *Environment and Planning A* 46 (5): 1160–80.

Goedhuys, M., and L. Sleuwaegen. 2010. "High-Growth Entrepreneurial Firms in Africa: A Quantile Regression Approach." *Small Business Economics* 34 (1): 31–51.

Green, V. 2003. *Emotional Development in Psychoanalysis, Attachment Theory, and Neuroscience: Creating Connections*. Hove, East Sussex and New York: Brunner-Routledge.

Hampwaye, G., and S. Jeppesen. 2014. "The Role of State- Business Relations in the Performance of Zambia's Food Processing Sub-Sector." *Bulletin of Geography: Socio-Economic Series* 26: 83–92.

Hart-Landsberg, M. 2013. *Capitalist Globalization: Consequences, Resistance and Alternatives*. New York: Monthly Review Press.

Hauge, J. 2016. "Africa's Industrial Policy Challenge Amid the Expansion of Global Value Chains: Industrial Policy in Ethiopia Since 2002." In *Africa's Turn to Industrialize? Shifting Global Value Chains, Industrial Policy and African Development*. London: School of Economics and Political Science.

Henley, D. 2015. *Asia-Africa Development Divergence: A Question of Intent*. London: Zed Books.

Hidalgo, C., B. Klinger, A. Barabasi, and R. Hausmann. 2007. "The Product Space Conditions the Development of Nations." *Science* 317 (5837): 482–7.

Himbara, D. 1994. *Kenyan Capitalists, the State, and Development*. Boulder, CO: Lynne Rienner Publishers.

———. 2016. *Kagame's Economic Mirage*. CreateSpace Independent Publishing Platform.

Hirschman, A. 1958. *The Strategy of Economic Development*. New York: Norton.

Hutchful, E. 2002. *Ghana's Adjustment Experience: The Paradox of Reform*. Oxford: James Currey.

Ibeh, K. 2015. "Rising Africa and Its Nascent Multinational Corporations." In *The Changing Dynamics of International Business in Africa*, eds. I. Adeleye, K. Ibeh, A. Kinoti and L. White. Basignstoke and New York: Palgrave Macmillan.

Jackson, R., and C. Rosberg. 1982. *Personal Rule in Black Africa: Prince, Autocrat, Prophet, Tyrant*. Berkeley, LA: University of California Press.

Jessop, B. 2002. *The Future of the Capitalist State*. Cambridge: Polity.

Johnson, C. 1982. *MITI and the Japanese Miracle: The Growth of Industrial Policy, 1925–1975*. Stanford, CA: Stanford University Press.

Kaplinksy, R., and M. Morris. 2016. "Thinning and Thickening: Productive Sector Policies in the Era of Global Value Chains." *European Journal of Development Research* 28: 625. doi:10.1057/ejdr.2015.29.

Kelsall, T., *et al*. 2013. *Business, Politics and the State in Africa: Challenging the Orthodoxies of Growth and Transformation*. London: Zed Books.

Khan, M. 2013. "Political Settlements and the Design of Technology Policy." In *The Industrial Policy Revolution II. Africa in the 21st Century*, eds. J. Stiglitz, J. Yifu Lin and E. Patel. London: Palgrave Macmillan, pp. 243–80.

Kissinger, H. 2011. *On China*. London: Allen Lane.

Lee, S., J. Wainwright, and J. Glassman. 2017. "Geopolitical Economy and the Production of Territory: The Case of US-China Geopolitical-Economic Competition in Asia." *Environment and Planning A*. https://doi.org/10.1177/0308518X17701727.

Leftwich, A. 2000. *States of Development: On the Primacy of Politics in Development*. Oxford: Polity.

Lensink, R. 1996. *Structural Adjustment in Sub-Saharan Africa*. London and New York: Longman.

Lieberthal, K. 1992. "Introduction: The Fragmented Authoritarian Model and Its Limitations." In *Bureaucracy, Politics, and Decision Making in Post-Mao China*, eds. K. Lieberthal and D.M. Lampton. Berkeley, LA and London: University of California Press.

Mamdani, M. 1996. *Citizen and Subject: Contemporary Africa and the Legacy of Late Colonialism*. Princeton, NJ and Chichester: Princeton University Press.

Mann, L., and M. Graham. 2016. "The Domestic Turn: Business Process Outsourcing and the Growing Automation of Kenyan Organisations." *Journal of Development Studies* 52 (4): 530–48.

Marais, H. 1998. *South Africa: Limits to Change: The Political Economy of Transition*. London: Zed Books.

Marshall, T. 2015. *Prisoners of Geography: Ten Maps That Tell You Everything You Need to Know About Global Politics*. London: Elliot and Thompson.

Martin, M. 1993. "Neither Phoenix Nor Icarus: Negotiating Economic Reform in Ghana and Zambia 1983–92." In *Hemmed in: Responses to Africa's Economic Decline*, eds. T.M. Callaghy and J. Ravenhill. New York and Chichester: Columbia University Press.

Mazzucato, M. 2011. *The Entrepreneurial State*. London: Demos.

Migdal, J. 1988. *Strong Societies and Weak States: State-Society Relations and State Capabilities in the Third World*. Princeton, NJ: Princeton University Press.

———. 2001. *State in Society: Studying How States and Societies Transform and Constitute One Another*. Cambridge: Cambridge University Press.

Moghalu, K. 2014. *Emerging Africa: How the Global Economy's "Last Frontier" Can Prosper and Matter*. London: Penguin Books.

Mohan, G. 2000. *Structural Adjustment: Theory, Practice and Impacts*. London: Routledge.

Moore, D. 1999. "'Sail on, O Ship of State': Neo-Liberalism, Globalisation and the Governance of Africa." *Journal of Peasant Studies* 27 (1): 61–96.

Moore, J. 2015. *Capitalism in the Web of Life: Ecology and the Accumulation of Capital*. London: Verso.

Morris, M., L. Plank, and C. Staritz. 2015. "Regionalism, end Markets and Ownership Matter: Shifting Dynamics in the Apparel Export Industry in Sub Saharan Africa." *Environment and Planning A* 48 (7): 1244–65.

Moseley, W. 2014. "Structured Transformation and Natural Resources Management in Africa." In *Managing Africa's Natural Resources: Capacities for Development*, eds. K. Hanson, C. D'Alessandro and F. Owusu. Basingstoke and New York: Palgrave Macmillan.

Muchie, M., P. Gammeltoft, and B. Lundvall, (eds.) 2003. *Putting Africa First: The Making of African Innovation Systems*. Aalborg, Denmark: Aalborg University Press.

Myrdal, G. 1968. *Asian Drama: An Inquiry into the Poverty of Nations*. London: Allen Lane.

Newman, C., J. Page, J. Rand, A. Shimeles, M. Soderbom, and F. Tarp. 2016. *Made in Africa: Learning to Compete in Industry*. Washington, DC: Brookings.

Noman, A., and J. Stiglitz. 2015. *Industrial Policy and Economic Transformation in Africa*. New York: Columbia University Press.

Nsehe, M. 2012. *Africa's Most Successful Women: Bethlehem Tilahun Alemu. Forbes*. cited. www.forbes.com/sites/mfonobongnsehe/2012/01/05/africas-most-successful-women-bethlehem-tilahun-alemu/#2309f0c811bc.

Oluwatobi, S., U. Efobi, I. Olurinola, and P. Alege. 2015. "Innovation in Africa: Why Institutions Matter." *South African Journal of Economics* 83 (3): 390–410.

Öniş, Z. 1991. "The Logic of the Developmental State." *Comparative Politics* 24 (1): 109–26.

Ouma, S. 2015. *Assembling Export Markets: The Making and Unmaking of Global Food Connections in West Africa*. Malden and Chichester: Wiley-Blackwell.

———. 2017. *Africapitalism: A Critical Commentary*. Mimeo.

Peiffer, C., and P. Englebert. 2012. "Extraversion, Vulnerability to Donors, and Political Liberalization in Africa." *African Affairs* 111 (444): 355–78.

Pempel, T. 1999. "The Developmental Regime in a Changing World Economy." In *The Developmental State*, ed. M. Woo-Cumings. Ithaca and London: Cornell University Press.

Oqubay, A. 2015. *Made in Africa: Industrial Policy in Ethiopia*. Oxford and New York: Oxford University Press.

Reno, W. 1998. *Warlord Politics and African States*. Boulder, CO: Lynne Rienner Publishers.

Ring, D. 2018. "Rwanda's Contested Model: Economic Rents, Development and Stability." http://roape.net/2018/06/12/rwandas-contested-model-economic-rents-development and stability/.

Rodrik, D., A. Subramanian and F. Trebbi. 2004. *Institutions Rule: The Primacy of Institutions Over Geography and Integration in Economic Development*. http://citeseerx.ist.psu.edu/viewdoc/download?doi=10.1.1.145.3355&rep=rep1&type=pdf.

Rolfe, R., A. Perri, and D. Woodward. 2015. "Patterns and Determinants of Intra-African Foreign Direct Investment." In *The Changing Dynamics of International Business in Africa*, eds. I. Adeleye, K. Ibeh, A. Kinoti and L. White. Basingstoke and New York: Palgrave Macmillan.

Rosenstein-Rodan, P. 1943. "Problems of Industrialisation of Eastern and South-Eastern Europe." *Economic Journal* 53 (210/1): 202–11.

Rotberg, R. 2013. *Africa Emerges: Consummate Challenges, Abundant Opportunities*. Cambridge: Polity.
Samatar, A., and A.I. Samatar. 1987. "The Material Roots of the Suspended African State – Arguments from Somalia." *Journal of Modern African Studies* 25 (4): 669–90.
Selwyn, B. 2014. *The Global Development Crisis*. Cambridge: Polity.
Singh, J.N., and J.S. Ovadia. 2018. "The Theory and Practice of Building Developmental States in the Global South." *Third World Quarterly* 39 (6): 1033–55.
Spruyt, H. 1996. *The Sovereign State and Its Competitors*. New Haven: Princeton University Press.
Staritz, C., L. Plank, and M. Morris. 2016. *Global Value Chains, Industrial Policy, and Sustainable Development – Ethiopia's Apparel Export Sector*. Geneva: International Centre for Trade and Development.
Stavrianos, L. 1981. *Global Rift: The Third World Comes of Age*. New York: Morrow.
Stein, H. 1992. "Deindustrialization, Adjustment, the World-Bank and the IMF in Africa." *World Development* 20 (1): 83–95.
Stiglitz, J., J. Lin, and E. Patel. 2013. *The Industrial Policy Revolution II: Africa in the 21st Century*. Basingstoke and New York: Palgrave Macmillan.
Storper, M. 1997. *The Regional World: Territorial Development in a Global Economy*. New York and London: Guilford.
Strange, S. 1996. *The Retreat of the State: The Diffusion of Power in the World Economy*. Cambridge: Cambridge University Press.
Swainson, N. 1980. *The Development of Corporate Capitalism in Kenya, 1918–77*. London: Heinemann Educational.
Tendler, J. 1997. *Good Government in the Tropics*. Baltimore, MD: Johns Hopkins University Press.
Thakkar, A. 2015. *The Lion Awakes: Adventures in Africa's Economic Miracle*. New York: St. Martin's Press.
Tomkinson, J. 2016. *Beyond Faith and Fatalism in Development Discourse: Global Conditions and National Development Prospects in Ethiopia*. Paper presented at the International Initiative for the Promotion of Political Economy Conference. Lisbon, September.
Tvedten, K., M. Hansen, and S. Jeppesen. 2014. "Understanding the Rise of African Business: In Search of Business Perspectives on African Enterprise Development." *African Journal of Economic and Management Studies* 5 (3): 249–68.
United Nations Conference on Trade and Development. 2012a. *Structural Transformation and Sustainable Development in Africa: Economic Development in Africa Report 2012*. Geneva: United Nations Conference on Trade and Development.
———. 2012b. *World Investment Report 2011: Non-Equity Modes of International Production and Development*. Geneva: United Nations Conference on Trade and Development.
United Nations Economic Commission for Africa. 2016. *Transformative Industrial Policy in Africa*. Addis Ababa: UNECA.
Urassa, G. 2014. "The Effect of the Regulatory Framework on the Competitiveness of the Dairy Sector in Tanzania." *International Journal of Public Sector Management* 27 (4): 296–305.
Van de Walle, N. 2001. *African Economies and the Politics of Permanent Crisis, 1979–1999*. Cambridge: Cambridge University Press.
Wade, R. 1990. *Governing the Market: Economic Theory and the Role of Government in East Asian Industrialization*. Princeton, NJ and Oxford: Princeton University Press.
Weiss, L. 1998. *The Myth of the Powerless State: Governing the Economy in a Global Era*. Cambridge: Polity.
Whitfield, L., O. Therkildsen, L. Buur and A. M. Kjær. 2015. *The Politics of African Industrial Policy: A Comparative Perspective*. Cambridge and New York: Cambridge University Press.
Woo-Cumings, M. 1999. *The Developmental State*. Ithaca and London: Cornell University Press.
Yeung, H. 2016. *Strategic Coupling: East Asian Industrial Transformation in the New Global Economy*. Ithaca and London: Cornell University Press.

CHAPTER

6

Deconstructing development

Post-modernism and decoloniality

> Development was – and continues to be for the most part – a top-down ethnocentric, and technocratic approach that treats people and cultures as abstract concepts, statistical figures to be moved up and down in the charts of 'progress'. . . . It comes as no surprise that development became a force so destructive to third world cultures, ironically in the name of people's interests.
>
> – (Escobar 1999, 382)

> The idea of development stands like a ruin in the intellectual landscape . . . the time is right to write its obituary.
>
> – (Sachs 1992, 1)

Is development a Trojan horse or a chimera? Some argue that the idea of development is in reality an imperial one which serves the interests of powerful states and actors in the international political economy, often to the disadvantage of those it is meant to benefit. Post-development theory first emerged in the 1980s and was then more fully elaborated in the 1990s. Its origins lie both in philosophical currents around post-modernism and post-structuralism and the concrete practices of people in the Global South organising against imposed hegemonic visions of development – what has sometimes been called developmentality (Lie 2015). This idea draws on Foucault's idea of governmentality, which is how people control themselves (their mentality) based on what they feel are authority structures in which they are enveloped (the governance part of governmentality). Foucault referred to governmentality as "the conduct of conduct". Developmentality, then, is performing certain sets of development policies, interventions and practices in accordance with the perceived wishes of powerful international development actors.

(Post)modernity

A good entry point to this discussion is perhaps the idea of post-modernity. "Post" obviously implies after something in terms of time, but what is "modernity"? According to

Comaroff and Comaroff (2012) modernity or being modern is a way of being in the world. According to them,

> modernity refers to an orientation to being-in-the-world, to a variably construed and variably inhabited *Weltanschauung* [worldview], to a concept of the person as self-actualizing subject, to an ideal of humanity as species-being, to a vision of history as a progressive, man-made construction, to an ideology of improvement through the accumulation of knowledge and technical skill, to the pursuit of justice by means of rational governance, to a restless impulse toward innovation whose very iconoclasm bring a hunger for things eternal.
>
> (9)

It would include things such as ideas originating from the European Enlightenment, such as a belief in science and that history is characterised by progress, and a shift away from more static ancient and medieval worldviews, such as the "chain of being". This theory posited that all matter and life were composed of a chain of being from God and angels, through kings, through the different social orders and ending with animals and minerals (Lovejoy 1936). As the king was at the pinnacle of the chain, at least in terms of human beings, they could decide who should live and die in the rest of the chain – the so-called "divine right of kings" to decide matters on earth given to them by God. Modernity then is associated with new forms of state and governmentality.

Enlightenment thinking disputed the existence of such structures and inspired political revolutions in France and United States, for example, with the US Declaration of Independence claiming that "all *men* are created equal" (emphasis added), excluding women, and initially the voting franchise was largely restricted to so-called "white" property-owning men. Modernity, then, is a belief system focussed on the power of science, technology and "rational" organisation of society. There are various streams of modernism, in art and architecture, for example. Perhaps the most famous modernist architect and town planner was Le Corbusier, whose work celebrated technology through high-rise housing developments.

As discussed earlier, the 1970s were a time of profound economic and political dislocation around the world. The Fordist compromise between big business, so-called big labour (mass trade unions) and the state began to unravel in the North Atlantic economies. Charles Jencks dated the end of modernist architecture to the destruction of the Pruitt-Igoe housing complex in St. Louis in 1972 (Harvey 1990). As noted earlier, President Nixon had taken the dollar off the gold standard the previous year.

The well-known geographer, David Harvey (1989, 3–4), wrote of Jonathan Raban's book *Soft City*:

> To the thesis that the city was falling victim to a rationalised and automated systems of mass production and mass consumption of material goods, Raban replied that it was in practice mainly about the production of signs and images. He rejected the thesis of a city tightly stratified by occupation and class, depicting instead a wide-spread individualism and entrepreneurialism in which the marks of social distinction were broadly conferred by possessions and appearances. To the

> supposed domination of rational planning . . . Raban opposed the image of the city as an 'encyclopaedia' or 'emporium of styles' in which all sense of hierarchy or even homogeneity of values was in the course of dissolution. . . . The city was more like a theatre, a series of stages upon which individuals could work their own distinctive magic while performing a multiplicity of roles.

Post-modernism disputes the precepts of modernism. It places less emphasis on social structures, such as class, and gives greater weight to individuals and their perceptions and beliefs. Post-modernism is suspicious of the idea of "truth" and investigates how certain things come to be taken to be true or become true as a result of powerful actors constructing discourses around them – "regimes of truth".

In his book *The Condition of Post-Modernity* David Harvey argues that the move towards post-modernism philosophically coincided with the shift away from Fordism to the dislocations of the new regime of capitalist "flexible accumulation". We can see these ideas around individualism and flexibility flowing into the rhetoric around the "sharing economy", such as AirBnB, for example, even though some of these practices can be highly exploitative and based on digital sweatshops (Ettlinger 2016).

Post-modernism was associated with rise of the "New Left" of the 1960s and 1970s which emphasised issues such as gay rights and women's rights and the environment. "Postmodern values emphasize self-expression instead of deference to authority and are tolerant of other groups and even regard exotic things and cultural diversity as stimulating and interesting, not threatening" (Inglehart 2000, 223). Post-modernism and post-structuralism seek to question and destabilise sometimes established categories. In language the technique of deconstruction sought to question or destabilise the relationship between words (signifiers) and things (signified). For example, love might be defined very differently by different people, so it has no single, stable, inter-subjective or established meaning for everyone.

Post-development

In the "development" field the adoption of post-modernism was associated with the rejection of the "Global Project" of modernisation being promoted by organisations such as the World Bank and IMF (Esteva and Prakash 1998). For Esteva and Prakash (1998, 2), for the majority of people in the Global South "modernization has always been for them, and will continue to be, a gulag that means certain destruction of their culture" and only through their own self-organisation and resistance can they escape its destructive effects. In their schema resistance doesn't necessarily mean active confrontation, as they note that one of Ghandi's most effective tactics against British rule in India was simply to ignore them – to delink from their control and to construct power autonomously. Aside from Ghandi, other influential thinkers who have influenced the development of post-development theory are Michel Foucault, with his ideas around how discourses, which are systems "of representation linked to relations of power with consequences on behaviour" (Ziai 2017, 66), create regimes of truth, as well as Ivan Illich who critiques mass compulsory education.

Development was seen as a discourse after President Truman announced a "program of development" for "underdeveloped areas" in his inaugural speech in 1949. The effect of this was to cast 2 billion people as underdeveloped, according to Esteva (1992 cited in Ziai 2017). In reality it is argued by post-development writers that the development discourse facilitated continued US and European intervention in ex-colonies in order to source raw materials and access markets and to be able to influence governments in those countries through aid, for example, for geopolitical and economic reasons. Through such discourse and development interventions many people in the Global South come to conceive of themselves as underdeveloped or poor (Ziai 2017).

Aid dependence has been extensively criticised in recent years, and Ivan Illich (1973) criticised Western modernity for making people dependent on the state. Writers in the post-development school argue for frugality, autonomy and sufficiency in contrast with neoclassical economics assumptions that people are profit and utility (happiness primarily through consumption) maximisers.

Probably the two most influential books in terms of post-development approaches have been by anthropologists: James Ferguson's (1990) *The Anti-Politics Machine: Depoliticization, and Bureaucratic Power in Lesotho* and Arturo Escobar's (1995) *Encountering Development: The Making and Unmaking of the Third World*, although Ferguson might not accept his inclusion in this school (Ziai 2017). Though Escobar's book focuses on Colombia and Ferguson's book on Lesotho in Southern Africa, both are concerned with the way in which development agencies develop discourses, or ways of thinking and talking about these places, which enable them to intervene in them.

As Vandana Shiva (1988) notes, poverty is something which is culturally perceived, although some people living without adequate food or shelter might disagree. Before the colonial encounter with European powers, indigenous people around the world had their own, often very culturally rich, societies. However, when Europeans arrived they perceived these people to be poor, which functioned ideologically to give them the perceived right to intervene and restructure these societies. In the north of Uganda theft of land by the British created overstocking of cattle by indigenous people on their reduced land areas. The British then forced destocking by making them sell "excess" cattle at depressed, administratively set prices, creating further poverty through a second theft – all in the name of conservation (Mamdani 1982). The longer-term results of this were famine.

A central concern of post-development work is the way in which people are constructed as backward, ignorant or poor and thereby disempowered and made into objects of intervention. This serves a variety of functions, such as allowing the claim that donor countries are helping those living in poverty, whereas in reality this type of objectification or "subjectification" (creation of subjects) may allow for exploitation, as in the case of Uganda described earlier.

Escobar traces the development discourse from President Truman's inaugural speech through to the time of writing in the 1990s. He notes, for example, that the United Nations' Department of Social and Economic Affairs wrote a report in 1951 which suggested the need for a complete restructuring of "underdeveloped" societies:

> There is a sense in which rapid economic progress is impossible without painful adjustments. Ancient philosophies have to be scrapped; old social institutions have

> to disintegrate; bonds of cast, creed and race have to burst; and large numbers of persons who cannot keep up with progress have to have their expectations of a comfortable life frustrated. Very few communities are willing to pay the full price of economic progress.
>
> (United Nations Department of Economic and Social Affairs 1951, 15)

This is an interesting quote for a number of reasons. First, it is totally in keeping with modernisation theory, which argued that the main obstacle to development was backward or traditional cultures. If these could be smashed, then "progress" towards development could be achieved. Second, aside from this cultural violence, development was also associated with other types of structural violence, particularly social exclusion and poverty for those who "could not keep up with progress".

In a sense the report was right – capitalist development is often a brutal process. For example, during the Industrial Revolution in Britain the majority of people got poorer (Polanyi 1944). However, it could be argued that this "short-term pain" was necessary for the long-term gain of generally increased and improved living standards. As noted earlier whether someone supports such an argument largely depends on their value system, social positionality and whether or not they believe certain people should have to be forced to pay the price for general social welfare improvement in the future. Similar arguments apply in relation to many development projects from large-scale dams to literacy. For example, should people be displaced from their ancestral lands to make way for large dams, which might provide cheap electricity and benefit urban dwellers and industrialisation? Who should have the power to make those decisions, and how should they be made? What should the relative weight of local communities in decision-making be relative to the "national" interest? Similar questions also apply in relation to the work of NGOs. As I noted in another book:

> NGO projects may also disrupt traditional social structures. For example, a NGO literacy project in the Philippines, by 'defining the women as lacking something – namely the ability to read and write – and by turning the project into a vehicle for education about modern values, the project contributed to an erosion of the status of older women and underlined a widening gap between educated professionals and peasant women' (Hilhorst 2003, 100). Also the intervention model which many NGOs use require them to acquire a lot of power, without too much 'interference' from local communities.
>
> (Carmody 2007, 55)

Depending on one's value set, it could be argued that it is local communities which should ultimately decide how their land and resources are used and that they have often been able to sustainably manage these for millennia (Santos 2014). On the other hand, it could be argued that industrialisation offers the prospect of a better future for the majority of people in a country and should therefore be pursued; although, there are obvious questions around the sustainability of current models of industry.

For Escobar (1995, 4), the promises of development were a lie, as it produced "massive underdevelopment and impoverishment, untold exploitation and oppression. The debt crisis, the Sahelian famine, increasing poverty, malnutrition, and violence are only the most pathetic signs of the failure of forty years of development". However, this is an extensive list, and not all the failures listed can be placed at the door of "development". For example, the debt crisis of the 1980s was produced by a variety of structural and conjunctural factors. It could be argued that the debt crisis primarily arose out of what Cowen and Shenton (1996) term "immanent development" – that is, the laws of motion of the global capitalism, rather than "imminent" development, which seeks to guide the process through human agency and policies. In a similar vein Hart (2001) distinguishes between what she calls "big D" (imminent) and "small d" (immanent) development.

A second problem with Escobar's claim is an empirical one. For example, life expectancy around the world has been increasing, reflective of generally better health systems and improved living conditions, although this is not the case for some places and regions. In his native Colombia, life expectancy, which is arguably a better indicator of general social wellbeing than economic growth, much of which may be captured by an elite, rose from 1960 to the mid-2010s and from the mid-50s to the mid-70s (Actualatix 2017). These gains have been mirrored in much of the rest of the world. In the 20 years from 1990 to 2010, life expectancy in Bangladesh rose from 59 to 69 (Radelet 2015).

Much "global progress" in recent decades, however, has been as a result of the remarkable industrialisation and economic transformation of China (Sachs 2015). This speaks to another issue or problem with Escobar's claim – the issue of scale. He claims that development has been an unmitigated disaster in the Third World; however, as described earlier, this has not been the case. In part it depends on how poverty is measured and defined. Using the World Bank (old) poverty line of US$1.25 a day, global poverty fell by 27 per cent from 1981 to 2005. However, if a less inhumane, higher line of US$2.50 is used, it increased by 13 per cent (Reedy and Pogge 2010 cited in Selwyn 2017).

In some cases and places "development" programmes have brought on disastrous socio-economic outcomes. For example, prior to economic liberalisation sponsored by the World Bank and IMF, Zimbabwe had quite an industrialised and diversified economy. However, through actions such as the allowing of second-hand clothing imports as part of its SAP, local industries were destroyed (Carmody 2001). There are many other examples of the ways in which development programmes have undermined lives and livelihoods; just as there are counter-examples of where they have improved them. This paradox is partly reflective of the ways in which development interventions interact with, shape and are shaped by broader patterns of combined and uneven development in the global capitalist economy. Of course, this does not invalidate some of the other claims that Escobar (1995, 9) makes, such as that like its colonial forebear, "development discourse is governed by the same principles; it has produced an extremely efficient apparatus for producing knowledge about, and the exercise of power over, the Third World".

James Ferguson's (1990) book was equally influential. In it he explored the ways in which the World Bank depictions of the socio-economy of Lesotho were completely at odds with the reality. The World Bank in its reports constructed the country as one dominated by subsistence agriculture that was largely unexposed to commercial markets or the money economy. However, the reality is that Lesotho has been part of the cash economy for centuries, partly driven by the huge importance of migrant remittances from mine workers in South Africa, which account for almost a third of the economy historically, although they have now fallen to 15 per cent (World Bank 2017d), partly as a result of South Africa restricting in-migration.

The purpose of constructing Lesotho in such a way was to allow the World Bank to find a role for itself in the management/discipline of that economy. If Basotho (the adjective associated with Lesotho) underdevelopment was seen to have been produced through interaction with the global capitalist and South African economies, that might have led to a radically different set of policy prescriptions and programmes.[1] However, constructing it as it did allowed the Bank to engage in development interventions, such as financing rural cattle markets. However,

> as Ferguson documents, almost all of these projects failed dismally to achieve their stated objectives, but criticism on those grounds is, he claims, pointless – because the projects did fulfill what he interprets as the (unstated) objective of justifying Bank involvement in Lesotho. Lesotho was constructed and defined as 'underdeveloped' in order to legitimize, and facilitate the exercise of western dominance over it.
> (Storey 2000, 41)

The World Bank had succeeded in constructing its version of the truth, or regime of truth, and partly succeeded in bringing it into being, at least as far as some people were concerned.

While not in the post-development school, another important book which examines development as a discourse is *The Will to Improve: Governmentality, Development and the Practice of Politics* by Tania Murray Li (2007). In this book Li examines how issues of poverty are discursively constructed in order to render them supposedly amenable to technical solutions or development projects. Li argues that rather than asking how poverty is created through unequal class relations and inequality, "rendering technical" these issues allows for certain types of development interventions which are often destined to fail. She argues that "programs that set out to improve the condition of the population in a deliberate manner have shaped Indonesian landscapes, livelihoods, and identities for almost two centuries" (1), beginning under colonialism.

This "will to improve", however, is parasitic on its own failings. That is, that as development projects often fail, they create the need for new development projects to alleviate poverty. In contrast to some other writers, however, Li argues that development is not a discourse designed to cloak its real intentions of continued resource, market and geopolitical exploitation. As she writes "I take seriously the proposition that the will to improve can be taken at its word. . . . Interests are part of the machine, but they are not its master term" (9). Of course, different elements of the state apparatus in donor countries may have different intentions.

Li posits that there are a number of contradictions in the development process between, for example, "the promotion of capitalist processes and the concern to improve the condition of the dispossessed" (31), when, in fact, it was capitalist processes which produced the dispossession in the first place through land grabbing and the creation of land markets. In the development projects in Indonesia which she studied, she found that "villagers resented being required to spend long hours and days in meetings drawing pictures and making themselves attractive to patrons, for which they received minimal material payback and no serious response to the fundamental problem of access to land" (277). Consequently

> practices of government limit the possibilities for engaging with the targets of improving schemes as political actors, fully capable of contestation and debate. They do this by inscribing a boundary that separates those who claim to know how others should live from those whose conduct is to be conducted.
>
> (282)

Post-post-development?

Post-development thinking has been subjected to sustained and substantial critique, although it has reinvented itself in other ways, discussed later under the rubric of decoloniality. Post-development approaches have sometimes been critiques for idealising, reifying or essentialising the idea of community, without paying attention to the ways in which communities are often fractured along class, ethnic or other lines, for example (Nederveen Pieterse 2000) and consequently the ways in which communities can be a source of violence.

A concomitant or related failing of post-development theory, as intimated earlier, is that it neglects positive development experiences and the fact that many popular struggles in the South are precisely around accessing development (Rangan 2004). Wishing development away does not obviate the needs for potable water or proper sanitation, although there may be alternative ways of achieving these through co-production of water between government and communities, rather than water privatisation, for example. "To ignore that desire is to romanticize the aspirations of many ordinary people – precisely the type of cultural imperialism post-development theorists claim to reject" (Storey 2000, 42). Furthermore, as Storey notes, social movements are not necessarily progressive, but may have strong reactionary elements, so to essentialise them is again misplaced.

Another of the flaws of post-development thinking is related to its foundations in post-modernism, according to some. Post-modernism challenges the idea of truth, à la science, and universalising ideas. However, this creates a paradox where the only truth is seen to be that there is no truth – a logically unsustainable position – only perceptions and discourses. However, the existence of discourses, such as a universalising development one, and perceptions are also truth claims. "The emphasis of much postmodern literature on playful, leisured, heterodoxical self-indulgence also has little to offer those who can still only aspire to safe drinking water, a roof which does not leak and the like" (Simon 1997, 186).

Another potentially dangerous aspect of post-developmentalism is that it shares with neoliberalism an anti-statist bias (Simon 1997). Calls for greater community participation may actually serve as a cover for the hollowing out of the state and a responsibilisation of communities for their own poverty, rather than examining the ways in which this might have been created through the practices of colonialism and unequal exchange (Emmanuel 1972), where low-value goods such as raw materials are exchanged for higher-value manufactured ones, for example. Beyond a celebration of social movements and NGOs, it is not clear what alternative post-development is offering (Simon 1997). The state has a vital role to play in promoting development, although its capacities and role will vary by context, depending on the nature of the state and the balance of social forces. However, post-development reifies and celebrates resistance over emancipation. It plays down class as an analytical category and consequently is populist in orientation (Nederveen Pieterse 2000).

Finally some have argued that development discourse is not in fact colonial, but represents a fundamental break with it because it contains "a firm assertion of people of all races to participate in global politics and lay claim to a globally defined standard of living" (Cooper 1997, 84 quoted in Ziai 2017, 78). However, others dispute this, arguing that the United Nations and World Bank official poverty line of US$1.90 a day is racist, as it would not be accepted as a reasonable poverty line in "developed" countries (Selwyn 2017).

Even some post-developmentalists have now accepted some of the critique. For example, Wolfgang Sachs, who is often considered to be one of the founders of this intellectual current, now argues that "delinking the desire for equity from economic growth (and linking it to new notions of well-being) and achieving a transition from economies based on fossil-fuel resources . . . are the cornerstones of the post-development age" (quoted in Ziai 2017, 74). Perhaps, somewhat ironically, this seems to be very close to what his namesake, Professor Jeffrey Sachs (2015) of Columbia University, argues for in terms of sustainable development. Jeffrey Sachs has also been involved in the production of the World Happiness Report. Thus, it would appear that some mainstream development theorists and practitioners and post-developmentalists have converged on the idea of quality of life in the context of sustainability as a worthy societal goal. Perhaps it is the outcome rather than the terminology which is ultimately more important.

Living post-development? Zapatismo and *buen vivir*

One of the social movements that most inspired post-development thinkers has been the Zapatistas in the southern Mexican, predominantly indigenous, state of Chiapas. This movement took over control of much of the state on the day that the North American Free Trade Agreement (NAFTA) came into effect in 1994. NAFTA was a free trade agreement between Canada, the United States and Mexico, which the Trump administration in the United States renegotiated and renamed after coming into office – it is now known as the USMCA (US, Mexico, Canada) Agreement. Kingsnorth (2003, 3) wrote:

> What may turn out to be the biggest political movement of the twenty-first century emerged from the rainforest remnants of southern Mexico on 1 January 1994,

> carried down the darkened, cobbled streets by 3,000 pairs of black leather boots at precisely thirty minutes past midnight. The owners of the boots carried rifles and the odd AK-47 or Uzi. . . . Three thousand faces, hidden by black woollen ski masks, bore the distinctive features of the Mayan Indians of Central America; a people outgunned, outcompeted, pillaged, slaughtered or simply passed over since the Spanish conquistadores first arrived on their shores in the sixteenth century. Now, half a millennium later, here in Chiapas, Mexico's poorest and southernmost state, 'the ones without faces, the ones without voices' had come to make the world listen.

According to one of the main leaders of the rebellion, Sub-Commandante Marcos, NAFTA was a death sentence for indigenous people in Mexico. There were two main reasons for this. The first was that maize production, as the main staple food, was central to many indigenous livelihoods. Heavily subsidised corn from the US was seen to represent a threat to these lifestyles and livelihoods by potentially driving indigenous producers to the wall.

> The inclusion of maize and beans in the NAFTA negotiations represented the final break with policies to protect small producers. In Chiapas, as in Mexico as a whole, over 80 percent of ejidatarios [small farmers] grow maize.
>
> (Harvey 1995, 194)

The proportion of imported food in Mexico's total increased dramatically in the four-year period after 1992 from 20 to 43 per cent (Shiva 1999).

The second main objection that the Zapatistas had to NAFTA had to do with land ownership. The Zapatistas took their name from Emiliano Zapata, who was a Mexican revolutionary and agrarian reformer. After the Mexican Revolution from 1910 to 1920 there was substantial land reform in the country, as many large *haciendas* (large estates) were broken up and an *ejido* system of communal or community land ownership and management was put in place.

Prior to the revolution landlessness had been a major problem, and the *ejido* system was seen to be one way of overcoming this by preventing or reducing private land sales. However, as NAFTA was in the offing, there were substantial reforms to Mexican land laws. As of 1991 *ejido* lands could be bought and sold legally or rented and used as loan collateral (Harvey 1995). This effective privatisation of *ejido* land again opened up the possibility of widespread land dispossession and landlessness. By 1910, the start of the Mexican Revolution, the indigenous population of Mexico had been deprived of 90 per cent of their land (Stephen 1998), and there were fears that such a violent process or appropriation would take place again. However, while there has been some loss of land, widespread landlessness has not been reported as a result of NAFTA, partly because the government also enacted a programme of certification of *ejido* rights, which enabled people to register their land.

In a sense the Zapatistas embody many of the precepts of post-development, with some considering their movement "the first post-modern revolution" (Kingsnorth 2003, 7).

They emphasise local autonomy over central government control. However, in a sense they also embodied some of the contradictions of post-development:

> On the one hand, the Zapatistas called for the democratization of the political system, reinforcing similar demands being made within Mexican society. But they were never able to make precise the meaning of their political project, besides the obvious condemnation of electoral fraud.
>
> (Castells 1997, 81)

Sub-Commandante Marcos was also a somewhat contradictory figure. He published conversations with a snail on the Internet. The snail could perhaps be taken to represent the viewpoint of less powerful, indigenous people. However, Marcos was not indigenous and was highly educated, having a master's degree from one of Mexico's most prestigious universities. There are a variety of theories as to why he was called sub-commander, with one being that the people were the commanders and hence he was only the sub-commander. However, the fact that he was very articulate and educated made him attractive to the global media, even though he was not of indigenous extraction.

While the Zapatistas achieved widespread publicity in international media, they have not proven to be the revolutionary force that some might have hoped they would become. While they have been successful in maintaining their organisation over decades, the Mexican state has not been fundamentally challenged. However, perhaps that was not the objective. Holloway (2002) argues that it is possible to "change the world without taking power" and gives the Zapatistas as example.

Around 200,000 Zapatistas live in 1,000 communities (Ziai 2017). Their economy is substantially based on collective, as opposed to individual, ownership and production, and they don't accept government subsidies in order to maintain their autonomy and independence. So the Zapatistas have been successful in crafting and implementing their own vision of society, even if it hasn't spread to other regions in the way that some might have anticipated. However, there have been other innovations in other relatively less wealthy southern states in Mexico, such as Ooxaca, which neighbours Chiapas and which claimed its own sovereignty, independent of the central government. It has also been a centre of the food sovereignty movement, which argues for localised production and control of food, rather than places being subject to the vagaries of the globalised agro-food production system, which is inherently unsustainable (Weis 2007).

Food is of central importance to the identity of indigenous communities in Mexico. For example,

> surveys of Zapotec indigenous households in the state of Oaxaca – an important center of corn genetic diversity – found that despite mean total production costs of more than 400% above the market cost of corn, families continued to plant and consume many traditional varieties instead of (or in addition to) purchasing corn, for reasons that include perceived higher quality, nutritional superiority, and cultural factors.
>
> (Chappell *et al.* 2013)

So-called *Homo economicus,* or "economic man", wouldn't approve.

Powerful food sovereignty movements have also arisen elsewhere in Latin America to protect and promote autonomy over food provisioning in the region. These are in contrast to the idea of food security, which is often associated with increasing food output, even though there is, by some estimates, twice as much food in the world than is needed to feed everyone adequately (Nally 2016). TNCs often promote the idea of food security and argue that this can be achieved by increasing output, whereas in reality the issue is that the current TNC-led global food regime distributes food very unevenly (Friedmann and McMichael 1987).

Food is central to another important idea which has gained traction in Latin America in recent decades – *buen vivir*, or good living. While there is no one definition of *buen vivir*, it is closely related to ideas around community, sufficiency and frugality – quite similar to ideas in ecological economics. According to Gudynas (2011) (cited in Ziai 2017) the term attempts to do away with, or perhaps does not recognise, the nature–society dualism. The idea of people being separate from nature can be considered one of the core features of Enlightenment epistemology. However, people can be considered part of nature, and some now write of socio-nature to communicate that. What constitutes nature or the natural is discursively constructed. The idea of *buen vivir* has been incorporated into the constitutions of Ecuador and Bolivia in the last decade, partly as a result of the power of indigenous movements.

Whereas the Zapatistas have not sought state power, Bolivia offers something of a counter-example where a social movement led by Evo Morales did come to power in 2006. The political party which he leads is called the Movement towards Socialism and arose from resistance to privatisation of water and gas resources, culminating in the so-called water and gas wars of 2000 and 2003.

Bolivia is the poorest country in South America. Life expectancy is approximately 69 years of age, having risen from 44 years in 1964. Like all of the countries of the region, there is a history of brutal colonisation and associated inequality (Galeano 2009). A large part of this inequality arises from unequal access and ownership of land and other resources, with European settlers or their descendants monopolising most of these, and indigenous people often suffering under feudal-style arrangements where they were effectively bonded to large land owners. According to an indigenous leader:

> First it is necessary to see the situation in which we lived – *empatronadas*, in a system of slavery, without rights to land, gripped only the hand of the patron. It's from there that our ancestors decided to organize themselves [in the late 1980s], first to consolidate or recover their territory, then to recover their freedom, and finally, to recover their cultural identity. Guaraní leader from Itika Guasu, April 21, 2009 quoted in Anthias.
>
> (2018, 19)

Bolivia is, however, a resource-rich country, with substantial gas deposits in particular. Conflicts over resources became acute in the early 2000s when there was an attempt to privatise the municipal water supply in the town of Cochabamba. The American multi-national Bechtel was given the contract to run the water supply, but in order to make it profitable, a law was passed which made it illegal to collect rainwater (Pilger 2002). This resulted in mass protests – the so-called "water wars". This then fed into the subsequent "gas wars" around who should exploit and benefit from this resource. These popular mobilisations fed into the election of Bolivia's first indigenous president – Evo Morales in 2006, who nationalised the country's gas reserves. However, despite his Movement towards Socialism declaring itself to be a post-neoliberal government, it has been critiqued for not fundamentally changing agrarian structures and class relations (Brabazon and Webber 2014) and continuing to allow multi-national exploitation of gas reserves, with some arguing that this is a form of "neo-extractivism", which largely neglects the development potential and needs of rural areas (North and Grinspun 2016). This was done partly in order to fund government services (Andreucci and Radhuberb 2017).

The fact that TNCs continue to be heavily involved in natural resource extraction in the country has led some to argue that there has not been a fundamental break with what is an exploitative system. Authoritarian tendencies have also emerged in the government, with Evo Morales effectively ignoring a defeat in a referendum to extend presidential term limits, with the Constitutional Court ruling in 2017 that he could run again, for a fourth term. As Lord Acton's adage goes, (unrestrained) "power tends to corrupt and absolute power corrupts absolutely". Ecuador, on the other hand, attempted to break with extractivism by trying to get richer countries to pay for oil *not* to be pumped out of Yasuní National Park. However, substantial funding was not forthcoming, and drilling commenced as of 2016 (Vidal 2016).

Post-colonialism and decoloniality

Post-colonialism or post-colonial studies is sometimes known as subaltern studies. A subaltern was a junior officer in the British army and literally means subordinate. The Subaltern Studies Group is a group of scholars of South Asia who are interested in post-colonial societies and the impacts which colonialism had on their formation and functioning. Given their highly critical stance towards colonialism, it is somewhat ironic that the group used a term from the British army to indicate someone of subordinate social standing or status, although they took the term from the Italian Marxist Antonio Gramsci who wrote about subaltern classes.

Perhaps the most famous of this group of scholars is Guyatri Chakravorti Spivak, who is a professor at Columbia University in the US. Her best-known work is an essay called "Can the Subaltern Speak?" (Spivak 1988). In some of my classes where I have students do a response paper, I tell them that they shouldn't write that the papers were hard or difficult to follow: the challenge, I say, is to understand and critique them. Many readers, including this one, will certainly find "Can the Subaltern Speak?" a challenging piece of work.

The book chapter essentially deals with issues of representation. In particular, it asks whether or not "white", "Western" critical scholars can accurately speak for those who they claim to represent – "subalterns" – people and classes of colour in the Third World. Her foil is a conversation or interchange between the two famous post-structuralist scholars Michel Foucault and Gilles Deleuze. She takes issues with the categories they use and argues that they, through their language, may actually be reproducing colonial discourses, even if that is not their intent. For example, she objects to the way the term "Maoism" (the philosophy expounded by former Chinese Communist Party leader Mao Tse-tung) is used and deployed, "which would be a harmless rhetorical banality were it not that the innocent appropriation of the proper name 'Maoism' for the eccentric phenomenon of French intellectual 'Maoism' . . . symptomatically renders 'Asia' transparent" (272). Thus, she is arguing that this is a claim to knowledge of, and thereby power over, the way in which Asia is discursively constructed, despite these post-structural scholars being critical of power and its operation. In common with other scholars writing after her, she is critical of the use of supposedly universal concepts such as class without contextualisation.

Spivak argues that by romanticising "the other" – the subaltern – intellectuals, perhaps unwittingly, replicate colonialist tropes and put responsibility on subalterns for their own liberation, thereby perhaps eliding their own responsibilities in the creation of injustice. Spivak draws an analogy with the British outlawing of the Hindu practice of sati, or wife burning, on a husband's funeral pyre. While this ban may have appeared progressive, it was used to justify British imperialism in India: contrasting British "civilisation" with Indian "barbarism", despite the fact that British rule in India was itself barbaric. For example, Britain oversaw the Bengal famine of 1943, when millions of people died as food was diverted to already well-supplied British troops elsewhere (Mukerjee 2010).

According to Das (1989, 312) subaltern studies "make an important point in establishing the centrality of the historical moment of rebellion in understanding the subalterns as subjects of their own histories". Post-colonial or subaltern studies is largely concerned with issues of representation and how just they are, and as such, its relationship to development studies may seem somewhat tangential. However, it is related to and feeds into other streams of thought around decoloniality in development, for example, although the two are distinct.

Mignolo (2007, 452) has argued that decoloniality is distinct from subaltern studies or post-colonialism, which he says is "a project of scholarly transformation" within academia. On the other hand the objective of decoloniality is thinking through how to delink from the "colonial matrix of power". This colonial matrix of power is traced back to the European Enlightenment and the practices of European colonialism which it is seen to have facilitated, even if many deny the interconnection between "scientific ways of thinking" and colonialism. In practice, however, the relationship between science and disciplines such as geography and anthropology and colonialism was a close one (Tilley 2011), as the roots of these subjects are colonial (Johnston and Sidaway 2004). Many colonial anthropologists spent their time measuring people's skulls, for example, in order to develop racial categories, or in some cases requested cut-off men's penises be sent back to Germany for analysis (Gordon 1998). The German government conducted a genocide against the Herero and Nama people in what is now Namibia between 1904 and 1907, with some arguing that this provided the template for the subsequent genocide against the Jews in Europe.

Again, perhaps one of the best-known decolonial theorists is Boaventura de Sousa Santos, who is a professor at the University of Coimbra in Portugal. His books *Epistemologies of the South: Justice Against Epistemicide* (2014) and *The End of the Cognitive Empire: The Coming of Age of Epistemologies of the South* (2018) contribute to the development of what is known as "Southern theory". This theory argues, as described earlier by Chakrabarty, that the extrapolation of European concepts and categories to other parts of the world, as if they were universal, is in fact inappropriate and a form of discursive violence or epistemicide – killing other ways of knowing. Rather, theory about the South should be developed using the Southern experience and lens, he argues.

For Santos, Global North and South are not geographical or social designations primarily, but political ones – that is, that one is either on the side of the oppressed or the oppressors. For him, there can be no justice without cognitive or epistemological justice. As noted earlier, epistemology refers to ways of knowing. So, for example, we could know a stream in terms of how it looks or makes us feel, or in a more scientific way about the amount of water that flows in it over a given time period. Thus, there are different ways of knowing and types of knowledge, which need not be mutually exclusive, but in practice some types of knowledge may get privileged over others. In particular, Santos argues that Western scientific knowledge, largely through colonial practices, devalued, displaced or overwhelmed other ways of knowing and being in the Global South. This is what he calls epistemicide. For him, there can be no justice without global cognitive justice.

Santos is one of the founders of the World Social Forum, which held its first meeting in Porto Alegre in Brazil in 2001. The location was significant because this is the city where participatory budgeting was initiated under Worker Party rule. Participatory budgeting is where citizens hold assemblies to decide on how government funds should be spent rather than this being decided by bureaucrats. The World Social Forum is a counter-point to, or perhaps in some sense the opposite of the WEF, which has met in Davos, Switzerland, every year since 1971. The WEF brings together politicians, policy makers and "captains of industry" and finance, or what Leslie Sklair (2001) calls the transnational capitalist class (TCC), which is composed of "four factions: those who own and control the major corporatons and their local affiliates, globalizing bureaucrats and politicians, globalizing professionals, and consumerist elites" (Sklair 2002, 144).

However, while it is often thought that social movements are oppositional to dominant social orders, this is not necessarily the case. Rather, the most powerful social movement in the world is arguably the TCC, which promotes neoliberalism globally (Murphy and Carmody 2015). As much "mainstream" media is owned by corporations and dependent on advertising for revenue from them, it often does not substantially challenge or helps produce this group's discourse and agenda (Herman and Chomsky 2008). The famous US investor, and one of the world's richest people, Warren Buffett famously said that there is a class war and that his class is winning. Given the structural power of transnational capital – that is, its ability to discipline states and shape governance ŕegimes to its own advantage through the threat of capital strikes (refusing to invest) or capital flight – this seems to be the case (Gill and Law 1989), at least in many places around the world. Southern theory has an important role to play in "unthinking" development as currently conceptualised. What comes afterwards is more difficult as people still need clean water, housing, jobs, etc. Can new forms of development, particularly so-called South–South cooperation, provide these more successfully and with less power inequality than that characterising Northern-led development to date? What role does or should ODA play in this?

Note

1 The fact that Lesotho had average gross capital formation of 47 per cent (1995–2005) – the highest rate in the world, also suggests this was far from just being a subsistence economy. This was largely as a result of the Lesotho Highlands Water Project. The project sells power to South Africa and originated under apartheid in that country, with World Bank funding. The Katse Dam, which was constructed as part of this, is on unsuitable geology, meaning that there are periodic earthquakes.

Further reading

Books

Escobar, A. 1995. *Encountering Development: The Making and Unmaking of the Third World: With a New Preface by the Author*. Princeton, NJ: Princeton University Press, 2012.

Li, T. 2007. *The Will to Improve: Governmentality, Development, and the Practice of Politics*. Durham, NC and London: Duke University Press and Chesham: Combined Academic [distributor].

Articles

Simon, D. 1997. "Development Reconsidered: New Directions in Development Thinking." *Geografiska Annaler: Series B, Human Geography* 79 (4): 183–201.

Storey, A. 2000. "Post-Development Theory: Romanticism and Pontius Pilate Politics." *Development* 43 (4): 40–6.

Websites

Mathews, S. 2010. "Post-Development Theory." *Oxford Research Encyclopedias*. http://internationalstudies.oxfordre.com/view/10.1093/acrefore/9780190846626.001.0001/acrefore-9780190846626-e-39.

Mercardo, J. 2017. "Buen Vivir: A New Era of Great Social Change." https://blog.pachamama.org/buen-vivir-new-era-great-social-change.

References

Actualatix. 2017. *Colombia: Life Expectancy (years)*. https://en.actualitix.com/country/col/colombia-life-expectancy.php.

Andreucci, D., and I. Radhuberb. 2017. "Limits to 'Counter-Neoliberal' Reform: Mining Expansion and the Marginalisation of Post-extractivist Forces in Evo Morales's Bolivia." *Geoforum* 84: 280–291.

Anthias, P. 2018. *Limits to Decolonization: Indigeneity, Territory and Hydrocarbon Politics in the Bolivian Chaco*. Ithaca and London: Cornell University Press.

Brabazon, H., and J.R. Webber. 2014. "Evo Morales and the MST in Bolivia: Continuities and Discontinuities in Agrarian Reform." *Journal of Agrarian Change* 14 (3): 435–65.

Carmody, P. 2001. *Tearing the Social Fabric: Neoliberalism, Deindustrialization, and the Crisis of Governance in Zimbabwe*. Portsmouth, NH: Heinemann.

———. 2007. *Neoliberalism, Civil Society and Security in Africa*. Basingstoke and New York: Palgrave Macmillan.

Castells, M. 1997. *The Information Age: Economy, Society and Culture: Volume II: The Power of Identity*. Oxford: Wiley-Blackwell.

Chappell, M.J., H. Wittman, C.M. Bacon, *et al*. 2013. "Food Sovereignty: An Alternative Paradigm for Poverty Reduction and Biodiversity Conservation in Latin America." *F1000Research* 2 (235).

Comaroff, J., and J. Comaroff. 2012. *Theory from the South, or, How Euro-America Is Evolving Toward Africa*. Boulder, CO: Paradigm Publishers.

Cowen, M., and R. Shenton. 1996. *Doctrines of Development*. London: Routledge.

Das, V. 1989. "Discussion: Subaltern as Perspective." In *Subaltern Studies VI: Writings on South Asian History and Society*, ed. R. Guha. New Delhi: Oxford University Press.

Emmanuel, A. 1972. *Unequal Exchange: A Study of the Imperialism of Trade*. New York: Monthly Review Press.

Escobar, A. 1995. *Encountering Development: The Making and Unmaking of the Third World: With a New Preface by the Author*. Princeton, NJ: Princeton University Press.

———. 1999. "The Invention of Development." *Current History* 98 (631): 382–6.

Esteva, G., and M. Prakash. 1998. *Grassroots Post-Modernism: Remaking the Soil of Cultures*. London: Zed Books.

Ettlinger, N. 2016. "The Governance of Crowdsourcing: Rationalities of the New Exploitation." *Environment and Planning A* 48 (11): 2162–80.

Ferguson, J. 1990. *The Anti-Politics Machine: "Development," Depoliticization, and Bureaucratic Power in Lesotho*. Cambridge: Cambridge University Press.

Friedmann, H., and P. McMichael. 1987. "Agriculture and the State System: The Rise and Fall of National Agricultures, 1870 to the Present." *Sociologia Ruralis* 29 (2): 93–117.

Galeano, E. 2009. *Open Veins of Latin America: Five Centuries of the Pillage of a Continent*. New edition. London: Serpent's Tail.

Gill, S.R., and D. Law. 1989. "Global Hegemony and the Structural Power of Capital." *International Studies Quarterly* 33 (4): 475–99.

Gordon, R. 1998. "The Rise of the Bushman Penis: Germans, Genitalia and Genocide." *African Studies Review* 57 (1): 27–54.

Hart, G. 2001. "Development Critiques in the 1990s: Culs de Sac and Promising Paths." *Progress in Human Geography* 25 (4): 649–58.

Harvey, D. 1989. *The Condition of Postmodernity: An Enquiry into the Origins of Cultural Change*. Oxford: Basil Blackwell.

———. 1990. "Flexible Accumulation Through Urbanization, Reflections on Postmodernism in the American City." *Perspecta-the Yale Architectural Journal* (26): 251–72.

Herman, E., and N. Chomsky. 2008. *Manufacturing Consent: The Political Economy of the Mass Media*. Anniversary edition with a new afterword by Edward S. Herman edition. London: Bodley Head.

Holloway, J. 2002. *Change the World Without Taking Power: The Meaning of Revolution Today*. London: Pluto.

Illich, I. 1973. *Deschooling Society*. Harmondsworth: Penguin Education.

Inglehart, R. 2000. "Globalization and Postmodern Values." *The Washington Quarterly* 23 (1), Winter: 215–28.

Johnston, R.J., and J.D. Sidaway, (eds.) 2004. *Geography and Geographers: Anglo-American Human Geography Since 1945*. 6th edition. London: Arnold.

Kingsnorth, P. 2003. *One no, Many Yeses: A Journey to the Heart of the Global Resistance Movement*. London: Free Press.

Li, T. 2007. *The Will to Improve: Governmentality, Development, and the Practice of Politics*. Durham, NC and London: Duke University Press; Chesham: Combined Academic [distributor].

Lie, J.H.S. 2015. "Developmentality: Indirect Governance in the World Bank-Uganda Partnership." *Third World Quarterly* 36 (4): 723–40.

Lovejoy, A. 1936. *The Great Chain of Being. A Study of the History of an Idea: The William James Lectures delivered at Harvard University, 1933 by Arthur O. Lovejoy*. Cambridge, MA: Harvard University Press.

Mamdani, M. 1982. "Karamoja: Colonial Roots of Famine in North-East Uganda." *Review of African Political Economy* 25: 66–73.

Mignolo, W.D. 2007. "Delinking: The Rhetoric of Modernity, the Logic of Coloniality and the Grammar of De-Coloniality." *Cultural Studies* 21 (2–3): 449–514.

Mukerjee, M. 2010. *Churchill's Secret War: The British Empire and the Ravaging of India During World War II*. New York: Basic Books.

Murphy, J.T., and P. Carmody. 2015. *Africa's Information Revolution: Technical Regimes and Production Networks in South Africa and Tanzania*. Chichester, West Sussex; Malden, MA: John Wiley & Sons Inc.

Nally, D. 2016. "Against Food Security: On Forms of Care and Fields of Violence." *Global Society* 30 (4): 558–82.

Nederveen Pieterse, J. 2000. *Development Theory: Deconstructions/Reconstructions*. London: Sage.

North, L.L., and R. Grinspun. 2016. "Neo-Extractivism and the New Latin American Developmentalism: The Missing Piece of Rural Transformation." *Third World Quarterly* 37 (8): 1483–504.

Pilger, J. 2002. *The New Rulers of the World*. London: Verso.

Polanyi, K. 1944. *The Great Transformation: The Political and Economic Origins of Our Time*. Boston: Beacon Books.

Radelet, S. 2015. *The Great Surge: The Ascent of the Developing World*. New York: Simon and Schuster.

Rangan, H. 2004. "From Chipko to Uttaranchal: Development, Environment, and Social Protest in the Garhwal Himalayas, India." In *Liberation Ecologies: Environment, Development, Social Movements*, eds. R. Peet and M. Watts. 2nd edition. London Routledge, pp. 205–26.

Reedy, S., and T. Pogge. 2010. "How Not to Count the Poor." In *Debates on the Measurement of Global Poverty*, eds. S. Anand, P. Segal and J. Stiglitz. Oxford and New York: Oxford University Press.

Sachs, J. 2015. *The Age of Sustainable Development*. New York: Colombia University Press.

Sachs, W. 1992. *The Development Dictionary: A Guide to Knowledge as Power*. London: Zed Books.

Santos, B. 2014. *Epistemologies of the South: Justice Against Epistemicide*. London and New York: Routledge.

Selwyn, B. 2017. *The Struggle for Development*. Cambridge and Malden, MA: Polity.

Shiva, V. 1988. *Staying Alive: Women, Ecology and Development*. London: Zed Books.

Simon, D. 1997. "Development Reconsidered: New Directions in Development Thinking." *Geografiska Annaler: Series B, Human Geography* 79 (4): 183–201.

Sklair, L. 2001. *The Transnational Capitalist Class*. Oxford: Wiley-Blackwell.

———. 2002. "Democracy and the Transnational Capitalist Class." *Annals of the American Academy of Political and Social Science* 581: 144–57.

Spivak, G. 1988. "Can the Subaltern Speak." In *Marxism and the Interpretation of Culture*, eds. C. Nelson and L. Grossberg. Basingstoke: Palgrave Macmillan.

Storey, A. 2000. "Post-Development Theory: Romanticism and Pontius Pilate Politics." *Development* 43 (4): 40–6.

Tilley, H. 2011. *Africa as a Living Laboratory: Empire, Development, and the Problem of Scientific Knowledge, 1870–1950*. Chicago, IL: University of Chicago Press.

United Nations Department of Economic and Social Affairs. 1951. *Measures for the Economic Development of Under-Developed Countries*. New York: Department of Economic and Social Affairs.

Vidal, J. 2016. "Ecuador Drills for Oil on Edge of Pristine Rainforest in Yasuni." *The Guardian*.

Weis, A. 2007. *The Global Food Economy: The Battle for the Future of Farming*. London: Zed Books.

World Bank. 2017. *Personal Remittances, Received (% of GDP)*.

Ziai, A. 2017. "Post-Development and Alternatives to Development." In *Introduction to International Development: Approaches, Actors, and Issues*, eds. P. A. Haslam, J. Schafer and P. Beaudet. Oxford: Oxford University Press.

CHAPTER

7 Aid, development and South–South cooperation?

> No country in the world can develop itself through foreign aid. That is a fact. To develop your economy is your job, you have to do it yourself.
>
> – (Chinese envoy to Malawi 2008 quoted in Masina 2008 in Brooks 2017, 249)

Dead aid?

In 2007 there was an illuminating exchange at a TED (Technology, Entertainment, Design) talk in the conference centre in Arusha, Tanzania. One of Africa's most well-known journalists, Andrew Mwenda, was giving a talk against aid to Africa. On the floodlit stage he asked rhetorically that if anyone knew of a country that had gotten rich on the basis of aid, could they please raise their hand. From the audience Bono, the lead singer from the Irish rock band U2, shouted up to him that Ireland had. Some people found Bono's intervention obnoxious or problematic. After all, there is a long and ignoble history of so-called "white" people telling Africans how to manage their affairs. (Ascribing importance or social characteristics to people based on their skin colour is a colonial practice, and some, such as Paul Gilroy, have argued that race should be abolished as an idea – hence "so-called 'white' people".) On the other hand, who has the right to decide who should and should not speak out about life-and-death issues? If we value free speech, perhaps we should be prepared to be challenged and made to feel uncomfortable, if necessary.

Aid is a topic that has aroused much passion and controversy in the last decade or so in particular. There are a number of well-known critics of aid, such as New York University professor William Easterly (2006, 2013) or P.T. Bauer (1984) of the London School of Economics, who received a knighthood for his work. Bauer famously argued that international aid was a system of giving money from poor people in rich countries to rich people in poor countries. However, nobody did as much to popularise anti-aid sentiment as the Zambian economist, Dambisa Moyo, with her 2009 book *Dead Aid*.

The premise of Moyo's book is that aid is the primary cause of Africa's underdevelopment. The mismatch between African needs and increased Western aid has arisen, she argues, because the most vocal global campaigners on Africa are Western, particularly Irish, celebrities. Moyo reviews the history of Western aid and then details what she sees to be its destructive effects: the creation of a dependency culture, enhanced opportunities for political corruption and distortionary macro-economic effects. Developmental failure, she argues, then increases the need for more aid. In a sense, it succeeds and self-perpetuates through failure.

According to Moyo, aid serves as a type of rent, where nothing has to be produced in order to receive it. Given that the private sector is weak in Africa, the best and the brightest "follow the money", but in its pursuit are corrupted. Aid, she argues, has served to prop up some of Africa's worst dictators, such as Robert Mugabe in Zimbabwe, at least until 2017, when he was removed in a palace coup to be replaced by his long-time right-hand man, "the Crocodile", Emerson Mnangagwa, who oversaw various campaigns of brutal repression during the Mugabe regime. This corruption, which aid fosters, she argues, in turn hampers private-sector development, which is the real way for Africa to escape poverty.

The macro-economic effects of aid are well known. When unearned foreign money, from oil, for example, enters an economy, dollars become more plentiful, meaning that less domestic currency, such as Zambian kwacha, are required to buy them. This is what economists call the law of supply and demand in operation. Prices adjust based on relative scarcity or abundance. In the absence of intervention by the central bank, inflows of foreign currency in the form of aid also makes the local currency more valuable. Consequently, exports become more expensive and imports cheaper, leading to trade balance problems, among other things.

Moyo argues that the solution to Africa's problems must come primarily from private-sector finance through accessing global capital markets for investment, encouraging remittances and microfinance for small businesses. This type of finance does not generate perverse incentives, as it operates on business principles, she argues. She cites some examples, such as the fact that the South African government used its (previously) good bond rating to underwrite an infrastructure investment fund for Africa or Kiva, a website based in California, which allows anyone to lend very small amounts to micro-entrepreneurs in Africa. Governments generally need to borrow to fund things such as infrastructure investment, and if they have a good enough credit score from the ratings agencies, they can do this by issuing government debt or bonds. Given political volatility and allegations of corruption in South Africa, its bonds were downgraded by Standard & Poor's, one of the main rating agencies, to "junk" (non-investment) status in 2017. In relation to microfinance, as discussed in more detail later, recent work commissioned by the United Kingdom's Department for International Development found that "microfinance can, in some cases, increase poverty, reduce levels of children's education and disempower women" (van Rooyen *et al.* 2012, 2259).

Moyo's book is written with conviction and a sound understanding of economics. However, the analysis places too much emphasis on the harmful impacts of aid, without examining its benefits – the children fed and inoculated or the small-scale hydro-electric dams built. While Africa grew poorer during the 1980s and the early 1990s, Moyo doesn't consider that it might have been worse in the absence of aid.

Moyo calls for most aid to be cut off to force self-reliance. If this happened, what would be the results in Ethiopia which, according to her, depends for 97 per cent of its government revenue on aid, although more realistic estimates suggest the figure is in the 50 to 60 per cent range? The author also ignores the substantial progress made in aid design and delivery recently, without presenting evidence that it does not reach the poor. Donors have sometimes designed systems to ensure effective tracking of funds, and in Uganda revenues received in local schools are now posted outside to ensure transparency. Indeed, although the author notes dramatic poverty reduction in Uganda in the 1990s, she fails to acknowledge that the country was heavily dependent on aid during that time.

Many of the issues that Moyo identifies are real and pressing. However, aid is not the primary cause of, or indeed solution to, Africa's problems. For her "good governance trumps all". But the global bargain whereby corrupt elites are supported by Western and other governments and companies, as long as they continue to allow resource exports and market access, is not addressed in the book. More fundamental changes in the distribution of power between African countries and the rich world and emerging powers, particularly in Asia, and between elites and more general populations, are required than simply cutting off aid. Aid can and does play a very useful role in international development, although it is also sometimes implicated in the production of poverty through some of the mechanisms Moyo describes and the conditions attached to it.

Aid conditionality and common pooling

Some of the conditionalities attached to IFI assistance have been described earlier, but aid often comes with conditions attached. One of the things that purportedly makes Chinese ODA different from Western sources is that it comes without conditionalities – "no strings attached". However, in some cases infrastructural loans from China come with the condition that much of the work must be undertaken by Chinese companies – the so-called Angola model or mode, where extensive loans were given in exchange for oil supply. This, then, represents a rescaling of conditionality to the project level, rather than its abolition.

When Gordon Brown was Chancellor of the Exchequer (finance minister) in the UK before becoming prime minister he used to talk about "doubling aid to halve poverty". This was an extremely naïve proposition and assumed that aid was the primary mechanism through which change comes about in the Global South. However, aid is implicated in the production of poverty, not just its reduction (Glennie 2008) through liberalising economic conditions often attached to it.

At the end of the 1990s there was widespread disillusionment with SAP policies. This led not only to their revision, through the implementation of PRSPs, but also wider discussions of the nature and effectives of ODA. Some Western governments had utilised policies of cross-conditionality with the IFIs, where aid was conditioned on having an SAP in place. However, the general disillusionment with these programmes led to changes in this. For example, when Tony Blair became prime minister in the UK in 1997, he set up a new Department for International Development (DfID), which has gone on to become perhaps the most high-profile aid ministry/agency in the world. Some even go so far as to claim that Britain is an "aid superpower" (Save the Children quoted in Quinn 2017).

Blair had a genuine interest in development, partly perhaps because his father had taught in Sierra Leone, but the setting up of DfID also served to placate old-style internationalists in the Labour Party who were uncomfortable with its general adoption of neoliberal policies. After leaving/being forced from office, when asked what her greatest achievement was Margaret Thatcher said, "Tony Blair and New Labour – we forced our opponents to change their minds".

Tony Blair was also relatively progressive in terms of how government policy should be formulated in the Global South, with the Commission for Africa arguing that it was up to governments themselves to design what their economic and social policies should be, rather than conditionalities being attached to financing. A general agreement emerged around this time amongst donors that development policies should be "government-owned" if they were to have any chance of success. It was argued that if policies were imposed from overseas that governments would have relatively little incentive to follow them through. This had implications for the way in which aid was to be delivered: in particular, through the advent of so-called direct budget support.

The theory of direct budget support and "common pooling" are related. The idea behind common pooling is that aid donors, rather than setting up discrete projects, such as hospitals, many of which might struggle once the initial funding runs out, should put their aid into a common pool to fund the government's operations. This was meant to imply that there wouldn't be multiple competing ways of delivering aid, but that it would centralise, support and help build government capacity to deliver its programmes. This is also known as direct budget support. Some donors such as the World Bank and DfID became strong proponents of this modality of aid delivery, whereas the American government remained sceptical, given the possibilities for corruption this might entail. These fears came to fruition in 2012:

> When Irish Aid suspended its entire assistance programme in Uganda. In October that year it was revealed that four million euros destined to help rebuild the country's war-torn northern region had been siphoned out to a personal account by the Office of the Prime Minister.
>
> (Yanguas 2018, 1)

In some cases the British government had attempted to prevent this possibility by insisting on the installation of a computer programme that would monitor government expenditure in Tanzania, for example. Interestingly, however, the fraud in Uganda was brought to light by that country's auditor general (Interview with diplomat, Kampala 2014). This modality of support remains in place, despite questions in the House of Commons in the UK around reported levels of fraud. Perhaps the importance attached to building government capabilities accounts for the support it continues to enjoy. In more recent years, there has been a return to more project-based approaches and an emphasis on evidence-based policy making associated with so-called randomised controlled trials.

Randomised controlled trials

An alternative to the approaches offered by Sachs (2005) around a "big push" around aid or Moyo's and Easterly's calls to reduce or eliminate aid are to be found in the work of Banerjee and Duflo (2012). They argue that it is not that aid is inherently good or bad, but that it needs to be targeted and delivered appropriately based on the experience of what is already working or might work. They argue that whether an aid intervention might work can be trialled on the basis of so-called randomised controlled trials (RCTs), which enable researchers to determine "what really works" in development. RCTs were initially developed in medicine in order to determine the effectiveness of drug treatments. How do they work?

RCTs work by identifying a target group or population which is to receive the intervention and then a control group who do not. In some cases there are more than one target group, who receive different types of interventions. After the target group has received the intervention, its impact can be measured relative to the control group. In order to ensure the scientific validity of the RCT, the key is to ensure the control group is a suitable comparator for the treatment group. This means that the characteristics of the control group should be as similar as possible to those of the treatment group before the intervention. Randomisation of the people, groups or firm to receive the treatment or to be in the control group is also important, as otherwise pre-existing differences might influence the measurement of the impact of treatment. When this approach is used, it is argued that it should be possible to attribute measured differences to the impact of the intervention.

In their book Banerjee and Duflo give examples of cases where RCTs can inform public policy. For example, in Mexico there is a well-known conditional cash transfer programme called *Progressa* which gives parents money as long as they keep their children in school. However, a World Bank study (Baird *et al.* 2009) found that conditionality did not seem to make a difference – parents will keep their children in school if they can afford to, so it was the payment alone which was important, not the conditionality.

RCTs also have a variety of problems, which have been well documented by Stein *et al.* (2018). One of the biggest issues is around the so-called "external validity". While a lot of effort is typically put into making sure the experiments are done properly to ensure "internal validity", what works in one context simply may not work in another or be "externally valid". Thus, "the use of RCTs within development economics has further entrenched a thin vision of economy which explicitly seeks to remove social, political, and cultural contexts" (13). As a result, they suffer from the same flaws, and arguably are part of conventional monoeconomics, which assumes the people everywhere are rational, self-interested economic maximisers (*Homo economicus*).

Sometimes people are paid to participate in RCTs, but this may result in different types of people being exposed to the intervention, as compared to those who might take it up if it were rolled out more generally by the government, for example (Barrett and Carter 2010 cited in Stein *et al.* 2018). Furthermore RCTs may identify if things happen at statistically significant levels, but not why things happen – that is, the causal channels through which interventions have impacts. However, this can be very important for policy, as policy makers typically want to understand why things happen to see if interventions can be made more effective.

Stein *et al.* (2018) also identify numerous other problems, such as spillover effects. So, for example, if a village is receiving aid through an RCT, the local government may divert spending from it. Their most serious critique is that RCTs sometimes harm people, even though one of the foundational principles in doing research is not to harm research subjects or participants. However, the way many RCTs are designed does not allow for informed consent of research subjects, and in some cases may create incentives which may harm their health status. Stein *et al.* cite one RCT in Malawi by Kohler and Thornton (2012) where participants were given cash incentives to maintain their HIV status for a year. However, the experiment produced perverse results, as men who received the payment were 12.3 per cent more likely to have vaginal sex than those who didn't, perhaps because more money opened up greater possibilities to pay for transactional sex.

According to Banerjee and Duflo (2012, 273):

> We also have no lever guaranteed to eradicate poverty, but once we accept that, time is on our side. Poverty has been with us for many thousands of years; if we have to wait another fifty or hundred years for the end of poverty, so be it.

However, at the beginning of their book they note that 9 million children die before their fifth birthday every year. I wonder if they and their parents would agree that time is on their side? As Jeffrey Sachs (2015, 482) rightly notes, "extreme poverty is the most urgent priority, because it is a matter of life and death for at least 1 billion people, and it is a struggle for survival in the here and now".

Microfinance

One of the interventions that Duflo and Banerjee discuss is so-called microfinance. Microfinance is spurred by the idea that one of the main barriers that poor people face is exclusion from financial markets. Because they are poor, they don't have collateral they can put up against loans, and because most of them don't have formal sector jobs with regular incomes, banks find it difficult to assess their likelihood of repayment, although new methods are being trialled in order to get around this, such as monitoring peoples' mobile phone usage to get a sense of their income and spending patterns. The basic idea behind microfinance is that once people living in poverty can access loans, they can start small businesses and lift themselves out of poverty. Sometimes aid is directed towards setting up microfinance operations.

There are different fads and fashions in international development which are sometimes seen as providing a "magic bullet" to problems of poverty. Microfinance was in vogue in the 1990s and early 2000s. In fact, Mohamed Yunus and the Grameen

Bank, which he founded in Bangladesh, are perhaps most associated with the idea and received the Nobel Peace Prize for their work in 2006. However, there have been a number of problems and critiques of microfinance and its methods and effectiveness.

One of these is the way it often relies on peer pressure in order to enforce loan repayment. For example, loans are often given to groups of women in the same village, and then it becomes their collective responsibility to repay, with shirking or non-repayment resulting in a variety of social sanctions. In some cases, however, where loans have not been repaid the Grameen Bank has resorted to tactics such as taking peoples' roofs off their houses – which might be considered a strange way to reduce poverty. Also Grameen Bank went to court to try and stop their borrowers from forming a union – again perhaps a strange way of promoting "empowerment" amongst poor women, who constitute the vast majority of its borrowers.

The actual experience of "poverty alleviation" promoted by micro-credit has been deeply problematic. According to Bond and Alam (2010),

> Grameen's origins are sourced to a discussion Yunus had with Sufiya Begum, a young mother who, he recalled, "was making a stool made of bamboo. She gets five taka from a business person to buy the bamboo and sells to him for five and a half taka, earning half a taka as her income for the day. She will never own five taka herself and her life will always be steeped into poverty. How about giving her a credit for five taka that she uses to buy the bamboo, sell her product in free market, earn a better profit and slowly pay back the loan?" . . . But what is the current situation in Jobra? Says Bateman, "It's still trapped in deep poverty, and now debt. And what is the response from Grameen Bank? All research in the village is now banned!" As for Begum, says Bateman, "she actually died in abject poverty in 1998 after all her many tiny income-generating projects came to nothing". The reason, Bateman argues, is simple: "It turns out that as more and more 'poverty-push' micro-enterprises were crowded into the same local economic space, the returns on each micro-enterprise began to fall dramatically. Starting a new trading business or a basket-making operation or driving a rickshaw required few skills and only a tiny amount of capital, but such a project generated very little income indeed because everyone else was pretty much already doing exactly the same things in order to survive."

Thus the same issue of the fallacy of composition that was noted earlier comes into play: that is, what might be rational for an individual or a country becomes irrational if the market is saturated and prices for what is being produced go down substantially.

Microfinance appealed to international "donors" because of its non-state, entrepreneurial (people supposedly pulling themselves up by their bootstraps) approach to poverty reduction. In fact, such was the attention given to microfinance as the "solution" to global poverty that it became the "domain equally of commercial banks, investment vehicles, and money markets" (Roy 2010, 5). However Roy questioned whether this "poverty capital" could serve the interests of the poor. In the end, it appears that the answer was no, as a systematic review for the United Kingdom's Department for International Development, quoted earlier, advised

> against the promotion of microfinance as a means to achieve the Millennium Development Goals – outcomes such as increased primary school enrollment do not increase micro-credit clients' ability to repay their loans and the diversion of finances to such long-term goals may lead to acute debt and increased poverty.
>
> (van Rooyen *et al.* 2012, 2260)

Systematic reviews are where researchers analyse the published literature, using a defined methodology, and see what the bulk of these studies suggest. Another study found that less than a quarter of microfinance projects funded by the World Bank and United Nations Development Programme were successful (Rosenberg 2006 cited in Roy 2010).

South–South cooperation

While it is only possible to scratch the surface of the literature and experience of different approaches to ODA in a short chapter such as this, one trend which has received increasing attention in recent years is so-called South–South or horizontal cooperation. According to its proponents, development cooperation between countries in the Global South is not marked by the same power hierarchies as North–South interactions and consequently can be more beneficial. This is what the idea of horizontality is meant to capture.

It is a truism to say that the world is undergoing a fundamental power shift or shifts. In recent decades the world has witnessed the rise of informational and informationalised capitalism. The global centre of economic dynamism and gravity has shifted eastwards, and emerging powers have dramatically increased their engagement in other parts of the developing world, such as Africa. Africa is also itself said to be "rising" (Mahajan 2009). What is the relationship between these phenomena, and how is the evolution of state-societies in the Global South being affected by these developments? Some of the other chapters in this book engage this question, but the question I want to focus on here is specifically around the area of so-called South–South cooperation (SSC), which is meant to be non-exploitative, in contrast to North–South relations.

The NAFC, centred in the West, and the rapid growth of a select group of developing countries, particularly (some of) the so-called BRICS powers (Brazil, Russia, India, China and South Africa) are two axes or dimensions of this rebalancing of the global geography of power. However, while the economic dimensions of this epochal shift have been explored through the Asian-Driver framework (e.g. Kaplinsky and Messner 2008) developed at a number of UK universities, the geopolitical dimensions have been less theorised. The Asian-Driver framework explores the ways in which emerging Asian economies, particularly China and India, are affecting development through a variety of channels such as aid, trade and investment. While many books and articles have explored the implications of the "rise" of China, relatively little work has been done which examines how the emergence and concentration between rising powers has affected global governance,[1] and the global aid regime in particular. One book which does that is Emma Mawdsley's impressive book *From Donors to Recipients: Emerging Powers and the Changing Development Landscape* (2012).

The book begins by taking us through the structure of the global aid regime, that is, the rules and institutions which govern how aid is disbursed around the world and debunking myths and misconceptions, such as that it is only recently that developing countries have instituted aid or development cooperation programmes. Mawdsley outlines the very substantial development cooperation provided over time by oil producers in the Arabian Gulf, Cuba or Brazil, for example. Cuba has a long history of sending medical professionals overseas. When Hugo Chávez was in power in Venezuela, he supplied Cuba with oil in exchange for doctors.

A variety of terms are used for emerging donors, such as non-DAC (Development Assistance Committee), but many of these are exclusionary or problematic, and consequently in order to do justice to the nuance and complexity of the relationships involved, Mawdsley settles on the term (re)emerging donors to describe them. (The DAC is a forum of the OECD rich country club, headquartered in Paris, where they discuss issues of aid or development cooperation.) The book describes the different modalities used by (re)emerging donors and how these differ or overlap with "traditional" Western approaches, and data are provided about the relative scale of (re)emerging donor assistance, which is thought to still account for a relatively small share of the total, although it is growing rapidly.

Mawdsley suggests that (re)emerging donor assistance has been more effective in achieving its goals of development effectiveness and rebalancing global power towards the South and East. She develops this argument using "gift theory". Whereas unreciprocated gifts or aid giving generates dependence and shame, the discourse of "mutual assistance" and "win-win" cooperation restores pride and (discursive) equality. While many Western commentators have described such rhetoric as an ideological camouflage for China's rapacious behaviour, in particular, the author shows that there is more to it than that, as development cooperation is about more than just material self-interest, even if that is an important part of the equation.

Mawdsley explores new development fora and institutions, such as the UN Development Cooperation Forum, which reviews trends and encourages coordination in ODA. Many of these new initiatives are both simultaneously progressive and exclusionary and have different impacts at different times. For example, after the NAFC the Western powers sought to better integrate emerging powers into global governance by giving greater importance to the G20 (Group of 20), which includes countries such as Brazil, Saudi Arabia and South Africa. In fact, it appeared to displace the G8 (Group of 8) as the primary forum for global economic discussion and coordination. The G8 was composed of the US, UK, France, Germany, Italy, Canada, Japan and Russia, which was subsequently suspended from the grouping after its annexation of the Crimean Peninsula from Ukraine in 2014. The G20 could have been seen as a democratisation of the international order. However, this new configuration excluded smaller, less powerful, often impoverished states, while arguably co-opting more powerful developing states into previous configurations of power. However, after the NAFC had passed, power was reconcentrated in the G7 (Stuenkel 2016).

Mawdsley also describes new modalities of development cooperation, such as the trilateral approach, where a developed country partners with an emerging power to deliver development assistance to a "recipient". There are a few points that are arguable in the book, such as that "aid must be understood primarily as a tool of foreign policy and thus subject to the same strategic calculations that are made in other areas" (27). Is it somewhat more complex than this? Does this vary by types of donors (Norway vs. the US)? Mawdsley challenges the conventional wisdom, such as donor harmonisation and coordination being necessarily a positive development. Rather, she argues that neoliberals should favour a "market" for aid which allows recipients to choose the most appropriate partners without the burdens associated with extensive (cross)conditionality. Brazilian aid is demand-led; that is, assistance must be requested in specific areas, rather than the "donors" deciding what areas they want to fund.

BRICS "cooperation" and developmental states in Africa?

There are many different types of SSC, and it is not possible to cover them here. Rather, I want to explore one axis or vector of engagement more closely. While the conditions surrounding the construction of developmental states were explored earlier, in this section I want to explore whether or not the BRICS engagement in Africa facilitates or hinders the creation of developmental states on the continent.

In a recent book (Carmody 2013) I outline the nature of the BRICS individual and collective engagements in Africa. I will not rehearse those arguments here, although I will draw on some of the concepts. Rather, my purpose here is a different one. I wish, in particular, to explore the limits, potentialities and impacts of the BRICS' stated policy of non-interference and the implications this has for African development into the future: in particular, whether or not the model of rapid economic growth and "strong states" of the BRICS is being transferred to Africa. In other words, are there "brics" (more statist, but fast growing economies) being built in Africa through their interaction with the most important emerging powers? How has rising BRICS engagement influenced the nature of African states and the content of their economic policies? Before doing this, however, it is necessary to briefly interrogate the nature of the BRICS themselves and the nature and scale of engagements with Africa.

Nature of the BRICS

There is a debate in the literature as to the usefulness of the concept of the BRICS grouping. In part this may be because of the etiology or origin of the term, which was originally coined by Jim O'Neill, a Goldman Sachs analyst, in 2001. It may also reflect an unease that the popularity of the term derives from its intuitive appeal (sounding as it does like bricks) rather than the structural characteristics of the participating political economies. As some of the BRICS have experienced economic problems recently, other acronyms which claim to better capture the most dynamic emerging economies in the world have emerged such as MINT (Mexico, Indonesia, Nigeria, Turkey) or the CIVETS (Colombia, Indonesia, Vietnam, Egypt, Turkey and South Africa). Some of these acronyms are also not just clever, but evocative. For example, a civet is a mammal which eats coffee berries and when they are excreted, they capture a high price as a result of this "processing". The implication is, potentially, that the CIVET countries also serve this value-added function in GPNs. However, these other groupings of countries cannot rival the scale and power resources of the BRICS countries, with the exception of South Africa, which is by far the least powerful in the BRICS grouping. What are some of the power resources of the BRICS?

China is the world's second-largest economy after the United States, and Brazil, Russia and India regularly feature in the indices of the world's ten biggest economies, depending on which agency is reporting the statistics. The scale of China's and India's populations are unrivalled, and they along with Russia are nuclear powers. After apartheid ended in 1994 South Africa became the first country in the world to voluntarily give up its nuclear weapons. China holds a permanent seat with a veto on the United Nations' Security Council, and Russia is the world's biggest energy exporter. That is not to say that the BRICS economies have not faced problems, such as slow or negative economic growth for Brazil and Russia in particular in recent years, or labour unrest and conflict in South Africa. However, as a grouping which is formally constituted, they hold substantial foreign exchange reserves and resources across different power fields.

In relation to Africa, some have argued that to focus on the BRICS engagement may be misplaced as some countries outside of the BRICS grouping have stronger economic ties than some of the members to Africa. Turkey's engagement on the continent has been particularly noteworthy and has been rapidly accelerating. However, while that country has opened seven new embassies in Africa recently (Harte 2012), the geography of Turkish economic engagement on the continent is very geographically specific. For example, the biggest economies in East Africa, Tanzania and Kenya, registered as having no Turkish investment as of 2012, whereas Egypt has almost a quarter of a billion dollars' worth and Mauritania US$66 million worth (Turkey 2014).

Scale and drivers of BRICS engagement

It is an oft-cited statistic that China is now Africa's largest single trading partner, with bilateral trade of approximately US$200 billion a year and also the largest source of new loans. However, others note that Africa's trade with the European Union is still greater than that and getting exact statistics on Africa is easy (Jerven 2013), but they are often unreliable.

Even the actions of the democratic members of the BRICS may further entrench authoritarian regimes in Africa. According to the Land Matrix database India was, in 2013, the fifth largest leasee and purchaser of land in Africa in a "new scramble" to produce food and agro-fuels in particular (Ottaviani 2013). One of the attractions of Ethiopia for Indian land investors is that it is an authoritarian country. Indian companies are desperate to source land overseas given that the country is losing a foot of groundwater a year, at least in the northern and central parts of the country (Rowden 2013).

South African and Brazilian companies are also substantial investors on the continent, and Russia is a major source of arms, although behind the United States in this respect. China was the single biggest supplier of arms to sub-Saharan Africa during 2006–2009 (Grimmett 2010).

The different BRICS have different drivers of engagement; however, all, barring Russia, are primarily driven by economic imperatives, although the precise dimensions of these vary. While natural resource access is a primary driver of Chinese engagement, Raphael Kaplinsky (2014) argues that market access is equally important. This contention is borne out by the trade statistics, as trade between China and Africa is roughly balanced. In contrast, the United States has historically run a substantial trade deficit with the continent, largely as a result of oil imports, although this has changed with the coming on stream of shale oil and gas in large volumes in the US.

Chinese-driven globalisation in Africa has a distinct sectorality, spatiality and temporality. In the literature there is a substantial debate about whether or not China is pursuing neoliberal policies. Indeed, recently there has also been a debate about what exactly constitutes neoliberalism, noted earlier. Leaving these debates aside, however, the structure of Chinese engagement in Africa is informed by the nature of state engagement and ownership in the Chinese economy.

The two sectors of the Chinese economy that remain most heavily state dominated are the natural resources and financial sectors. Both of these are critical for continued economic growth in China, as "dragon's head" state-owned natural resource companies, such as the Chinese Non-Ferrous Metals Corporation, are encouraged by that country's "go out" policy to source new reserves and supplies overseas and to grow, while in line with Keynes's dictum that finance should be primarily national, the banking sector remains primarily focussed on mobilising capital for domestic investment – although China is also the world's biggest exporter of financial capital given its massive trade surpluses (Fan Lim 2010). There is also still significant state ownership of the vital infrastructure and construction sectors.

The sectorality of Chinese engagement in Africa, led as it was by state-owned natural resource and infrastructure companies, has also led to a particular step-wise temporality to engagement. While the Angola model of resource for infrastructure swaps has sometimes been overstated in importance, it is a significant modality of engagement in Africa. It has also been controversial given loan conditionalities which sometimes require the contracting of Chinese companies to fulfil infrastructure requirements, often bringing Chinese labour with them. When these contracts are completed, some Chinese workers stay on and set up small businesses in trading, as restauranteurs or farmers, for example (French 2014). This deepens and further embeds processes of translocalisation and creates markets for other Chinese companies, such as banks. The sectorality, temporality and embeddedness of these processes serves to deepen interdependence between China and the African countries with which it has the most significant engagement.

The drivers and modalities of Indian engagement on the continent are different, even if they are heavily influenced by politics, or more accurately political economy. Whereas Chinese engagement is heavily state-directed and led, facilitated in part by that country's authoritarian political structure, in India that country's political system also influences the nature of engagement, even if in this case it is primarily private sector led. In China there is an authoritarian "social contract" whereby the populace largely seems to acquiesce to authoritarian rule in exchange for economic development and political stability. On the other hand, India is a vibrant – some might say raucous – democracy (Guha 2017). As I have argued elsewhere, this means that while there has been population displacement and dispossession in India in relation to the construction of major infrastructure projects, such as the Narmada Dam and SEZs, the doctrine of eminent domain cannot be used domestically for the appropriation of large tracts of land for agro-investments. (This doctrine asserts that the state has the right and ability to acquire private land for infrastructure, for example, when it is in the public interest.) This has resulted in the offshoring of this demand to countries in Africa in particular, given often falling agricultural yields in parts of India, land fragmentation and other issues. As Ian Taylor (2014) notes, India is also now setting aside financing for infrastructure and other projects in Africa to facilitate and deepen engagement.

Brazilian engagement in Africa is driven by a variety of imperatives, but foremost amongst these in the recent past was the need to try to sustain and reinvigorate the country's economy and thereby reproduce the class compromise that has underlay Worker Party rule in that country (Carmody 2013). The Temer government, which came to power in 2016, was more focussed on consolidating its own power domestically through repression, as noted by the UN High Commissioner for Refugees, for example, and consequently there was much less engagement in Africa.

South African engagement is also partly driven by political imperatives. Whereas apartheid was in part a "racial fix" to the blockages to capital accumulation in the mining sector, given often deep deposits and low-grade ores, and consequently the imperative of accessing subjugated "cheap" labour (Magubane 1996), the post-apartheid government has sought to use the rest of the continent as a spatial fix (Harvey 1982) to the current contradictions of capital accumulation in that country.

For Russia, while it has economic interests, Africa is primarily a space where "great power" status can be reasserted. The different nature of Russian engagement on the continent is reflective of the duality of human nature and state action. By way of illustration, the materialism of consumer capitalism helps constitute its opposite in Islamic fundamentalism, which attempts to reassert the primacy of brutal and authoritarian community over the anarchy and anomie of the market (Barber 1996). For Russia, the innate desire of people to be part of something bigger than themselves – in this case, a "great" nation – is a legacy of Russian and then Soviet power and imperialism, as well as its collapse. In Paul Kennedy's (1989) thesis great powers decline because they engage in military overstretch. There are signs that this might be a potential danger for Russia, in Ukraine and Syria, for example, even as President Obama in the US belittled it as a "regional power". For Vladamir Putin making Russia a great power again is a question of will, in the first instance, and of political and economic structures, in the second instance, which can be created through agency. He talks of a "vertical of power" (power being concentrated in his hands). The contours and terrain of engagement of Russia in Africa are shaped by this imperative.

Impacts of BRICS engagement

There are a variety of impacts of the BRICS engagement in Africa, which can be characterised as economic, political and social. The primary economic impact for much of the early 2000s, of Chinese demand in particular, was the commodity super-cycle and the reversal of the terms of trade between basic manufactures and primary goods. As companies such as Walmart engage in something called cost down-pricing, this has driven prices for consumer electronics and clothes, for example, down around the world. Cost down-pricing is where the buyer insists that the supplier (of clothes, for example) reduce the prices they charge each year of the contract.

Increased natural resource demand was at the heart of the African growth revival, although questions have been raised about the quality of this growth and its often enclavic nature (on the potential linkage and diversification effects from natural resources see Morris *et al.* 2012). High-commodity prices helped release the foreign exchange constraint on manufacturing in Africa, even if much of the sector is subject to competitive displacement pressures from more efficient and subsidised Chinese production. This resulted in a somewhat contradictory situation where the manufacturing sector's output is growing, even as its share of GDP contracts. This has important political implications, as natural resource dependence is often associated with poor governance, as governments can rely on rents from the sector rather than taxes from a more diversified economy.

There are also multiple social impacts of rising BRICS engagement in Africa. These include impacts on employment, incomes and social indicators arising from economic engagement and factors such as migration flows. For the purposes of this chapter, it is political impacts which are the primary focus: namely, whether or not brics are being built in Africa and whether new configurations are recasting state–society relations such that a new world order is in formation.

Building BRICS in Africa?

According to Ian Taylor (2014) emerging powers substantially differ from Western countries in their economic model, as there is a strong directive role for the state in the economy. As he puts it, "they . . . uphold a broadly statist approach to business that enables private enterprise and mercantile commerce. This state capitalism is increasingly coming to the fore, particularly in the aftermath of the global financial crisis" (2).

The West has recently undergone a prolonged and painful financial crisis. In contrast, much of Africa, until the end of the commodity super-cycle in 2014, experienced relatively more rapid economic growth. While there are questions about the environmental impact, sustainability and quality of Africa's economic growth, what accounted for these different economic trajectories? Part of the explanation lies in the growth of the BRICS economies and the investment, demand and aid they have generated for Africa. These economies are largely responsible for higher African economic growth through increased demand for primary commodities and investment in mining, infrastructure and other economic sectors. They are also changing the nature of globalisation and of the state–society formations with which they interact and influence.

As noted earlier, much investment from the BRICS, and from China in particular, is through state-owned corporations, such as the Chinese Non-Ferrous Metals Corporation. Rather than being driven by quarterly stock market returns, these companies have longer time horizons. The fact that the operations of these companies are influenced by strategic national planning also gives them a different character from Western corporations. Globalisation is often thought to be private sector led. This is not the case for some of the BRICS, where we are seeing a globalisation of state power. This globalisation of state power is also reflected in "recipient states" of Chinese and other BRICS investment.

The former Zambian minister of finance noted that if they privatised the mining industry in that country the government "would be able to access debt relief, and this was a huge carrot in front of us – like waving medicine in front of a dying woman. We had no option [but to go ahead]" (quoted in Action for Southern Africa 2007, 6). However, by 2011 a civil servant in the Ministry of Finance and National Planning was noting that "[a]t one point in time we had to agree with everything the donors said. Now we can disagree. The donors still act in a very coordinated manner but of late we have often disagreed with their policies" (quoted in Kragelund 2014, 156). This has meant that certain industries such as Zambia Railways and Zamtel, the telecommunications provider, have been taken back into state ownership.

Whereas Gillian Hart (2014) sees a dialectical tension between processes of economic denationalisation (globalisation) and the (largely) rhetoric of nationalism in South Africa, in the Zambian case this rhetoric was partly translated into policy, if slowly and incompletely. Consequently there in evidence of a tentative shift away from neoliberalism to neodevelopmentalism (Godio 2004), at least in parts of the continent. However, as the copper price fell and Zambia's debt-to-GDP ratio increased substantially, it started negotiations with the IMF about a new programme, which had stalled as of 2018.

One of the keys to China's economic success has been the development of SEZs (Stein 2012). As noted earlier, this model has also been exported to Africa and elsewhere, although there they are developed with Chinese rather than Western or overseas Chinese capital. In Africa these zones are mainly built as investments by SOEs, but their main tenants are small and medium-sized businesses, accounting for 85 per cent of the businesses in them (Tang and Zhang 2011 cited in Power *et al.* 2012). According to the Chinese ambassador to Zambia:

> We would also like to introduce mature Chinese enterprises with comparative advantages to Zambia to help address the country's over-reliance on import of consumer and manufactured goods. Therefore, the establishment of the Cooperation Zone can help both Zambia develop and mature Chinese industries redeploy and win more space of development at home.
>
> (quoted in Southern Weekly 2010 in Brautigam and Tang 2011, 31)

As Brautigam and Tang note, however, in contrast to China, none of the African zones have been designed to develop synergies with local technology institutes or universities, even if several of them have been developed by Chinese companies with experience of zone development at home. Also according to an analysis by Kolstad and Wiig (2011) the dominant motive for Chinese FDI in Africa is resource seeking, although there are also other motives, noted earlier. This fits with the dominant role that the Chinese state is playing in the development of these zones.

According to Cowaloosur (2014, 5):

> Regardless of whether the CSEZA [Chinese Special Economic Zone in Africa] companies are private or state-owned, Chinese government retains the reigns (sic)

> of the CSEZAs through its authority to select the developer, provide the funds and also because the CSEZAs are essentially bilateral state ventures between China and African countries.

A strong role for the state is thus in evidence. This is facilitated by the policies of non-interference propounded and promulgated by the BRICS powers.

The BRICS powers, both individually and collectively, have policies of "non-interference" in the domestic politics of their partner or client states, at least in terms of rhetoric. This is very attractive to incumbent African political elites, who tend to stay in power substantially longer than their counterparts in other world regions. The BRICS powers provide alternative sources of trade, investment, loans, aid and arms from traditional Western "partners". Whereas Western countries and institutions which they have historically controlled, such as the World Bank and IMF, have often insisted on free-market economic reforms and (sometimes) electoral democracy, or lately, "good governance" (Young 2012), this is not the case for the BRICS powers.

According to Obiarah (2007):

> China provides a powerful development model which urges economic growth before human rights, which has a number of possible effects on African development debates. First, African leaders can use this model to deny political rights to their people and rebuff efforts at building good governance and promoting democracy.
> (Quoted in Power *et al.* 2012, 181)

Furthermore as Patey (2014) notes, "global civil society" and human rights pressure on Western oil companies in Sudan, which forced their exit, paradoxically ended up improving the position of Asian oil companies, as they are often less susceptible to this.

The rise of the BRICS is also altering Western foreign policies in Africa. According to Verhoeven (2014, 58), by

> 2013 Western and Chinese dignitaries are both entertaining privileged ties with the same African elites. Diplomats consider the opposition irrelevant and pay attention almost exclusively to building relations with whoever is in power, a reversal [sic] of pre-1989 policies that African leaders have not failed to exploit.

Corkin (2013, 2) notes in relation to Angola that the "government has effectively managed to harness its relationship with China to bolster its position, both politically and economically". Thus, the power of extant political elites was solidified by the commodity super-cycle and declining aid dependence, for resource-rich countries at least.

However, the rhetoric of non-interference confronts a new reality. As levels of Chinese and other BRICS powers' investment in the continent have grown, they have necessarily become more embroiled in domestic politics in African countries. According to one Chinese ambassador in Africa, "of course we are increasingly involved in the politics of African countries, we are being pulled in, we have no choice. But do the Africans really want our input? I don't know, it is not clear" (quoted in Vehoeven 2014, 56). In any event, the extent to which "non-interference" was ever a reality is questionable. For example, arms sales to incumbent elites represent a form of extreme interference in internal affairs to maintain the status quo. The rhetorical attachment to "non-interference" is largely driven by political imperatives, partly to refute Western accusations of Chinese neo-colonialism in Africa (Corkin 2013). Also in practice, in Equatorial Guinea, for example, the ruling party receives aid from the Chinese Communist Party through the embassy in Malabo (Esteban 2010).

An Angolan academic has noted that Chinese involvement in Africa is "a good idea", as "China is the weapon that Africa can use to end western hegemony" (quoted in Corkin 2013, 41). However, the nature of economic interactions with China in particular has resulted in deindustrialisation across the continent, greater dependence on resource exports and consequently a diversification, rather than a transcendence, of dependency (Taylor 2014).

The answer to the question of whether brics are being built in Africa is therefore a mixed one. Some would dispute that there is a Chinese model of state–society relations, let alone a BRICS one (Taylor 2014). However, as noted earlier, "rising powers" are defined, in part, by the fact that there is a stronger role for the state in the economy – a "mixed economy". There was evidence during the commodity boom of movement towards more mixed economies in Africa, at least in parts of the continent. Policies of resource nationalism also continue, in Tanzania, for example, where the government in 2017 demanded US$190 billion in back taxes from the Acacia Mining Company (two centuries' worth of the gold producer's revenue) (Biesheuvel 2017).

The philosophy of non-interference amongst the BRICs is also one of the factors which accounts for the emergence of a new type of geopolitics. The growth of these emerging powers and their increasing projection of influence in Africa is reconfiguring global geopolitics, creating a new large-scale region noted earlier – a "South Space". This new macro-region is defined by a shared history of colonial exploitation and Western intervention, intensifying flows of trade, investment and aid, and increasingly, "hard shell" sovereignty, where governments in Africa are often able to resist calls to respect human rights, for example. At the same time the BRICS powers exercise increasing influence on the continent through aid, diplomacy and economic investment. By way of example, African countries which trade more with China are more likely to vote the same way as it in the United Nations General Assembly (Flores-Macías and Kreps 2013), at least on human rights–related votes (Carmody *et al.* 2018). This model of "influence without interference" represents a form of "dethority" or geogovernance. Geogovernance is the projection of "national" power across borders. However, the constituent countries of South Space are very different in terms of their economic policies and models of state–society relations.

The rise of the BRICS in Africa is a momentous event for the continent as its economic relations are reoriented from the West to the East and South. This has many progressive dimensions to it. Chinese and Indian demand for natural resources dramatically increased their prices to the benefit of many African economies, as these are often their primary exports. Chinese infrastructure for resource swaps, where new roads and railways are paid for in copper, for example, are less susceptible to corruption. Also the fact that the BRICS are less prescriptive about economic policy has opened up space for policy experimentation and learning based on African rather than Western conditions.

However, the rhetoric of "South–South Cooperation" and "win-win" globalisation disguises huge and often deepening power inequalities between African countries and China in particular, even if African state elites have been skilful in "balancing" between powers and managing extraversion for their own interests. As has been the case previously, political elites in Africa have been very skilful in converting and leveraging international relations to bolster their own regimes and authority. The BRICS powers have generally been happy to help them in this endeavour, as long as they are assured continued access to resources, land and markets. For example, the Chinese government reportedly helped Robert Mugabe's regime monitor Internet traffic and phone calls in advance of dubious and disturbing elections there.

According to Kurlantzick (2007) China prefers to maintain authoritarian regimes in power to prevent it becoming increasingly isolated in international affairs. While Adem (2010) argues that China has not completely shunned multilateralism, it prefers bilateralism in its relations with African states, as it is easier to influence/dominate them in this way. However, this does not mean that African political elites are supplicants to China or the other BRICS powers. As African countries often vote as a bloc at the United Nations, WTO and other fora, such as the International Olympic Committee, Bodomo (2009, 172) argues that "Africa sometimes wields more power than China in international settings", although there is no enforcement mechanism for votes at the United Nations General Assembly (UNGA). Furthermore, as gatekeepers of sovereignty and resource access, African political elites retain substantial power, although it is partly co-constituted through relations with other states, such as China.

In the literature recently there has been a heavy emphasis on the agency of African state elites in their dealings with China (Power *et al.* 2012). Corkin (2013) has shown, for example, how the Angolan state was able to negotiate that its loans from China would be repayed using the advantageous spot price for oil, rather than a lower contracted price. African state elites are acutely aware that the Western (near) monopoly on power and finance has been broken and strategically play off different foreign powers against one another. The dimension of structural power related to higher natural resource demand which African state elites are able to leverage are the effects of sunk costs. Once pipelines or oil facilities have been constructed, their embeddedness within the juridicially sovereign territories of African or other developing country states confers substantial bargaining power and leverage on political elites – the "obsolescing bargain". The American economist, Raymond Vernon, argued this was where the terms initially favoured the multi-national company, but this decreased through time as the scale of their investment increased. This is a phenomenon which is at play in relation to BRICS investments, but has also been in evidence in relation to the World Bank–funded Chad-Cameroon pipeline, for example (cf. Carmody 2010). According to Nossiter (2013):

> In Niger, government officials have fought a Chinese oil giant step by step, painfully undoing parts of a contract they call ruinous. In neighbouring Chad, they have been even more forceful, shutting down the Chinese and accusing them of gross environmental negligence.

The future of Africa and the BRICS

Some are pessimistic about the future growth of the BRICS (Sharma 2012). Hart-Landsberg (2013) notes that US trade deficits have fuelled growth in Latin America and Africa and that the Chinese economy is also dependent on these. Consequently, in the context of the NAFC crisis, he was pessimistic about the possibility for China to retain rapid economic growth, although this has continued. The enactment of substantial tariffs against Chinese exports to the US may affect this, however.

Growing economies in Africa, if they were sustained over the medium and longer terms, might mean the further enlargement of the middle and working classes, who have historically been better at making and keeping states more accountable (Rueschemeyer *et al.* 1991). In Africa, the struggle to ensure a fairer and more beneficial sharing of the results of a rebalancing, and increasingly Eastern-centred, global order has begun, as evidenced by the numerous organisations on the continent seeking justice for those displaced by land grabbing, for example. The future of the continent increasingly hinges on the outcomes of these struggles and the continent's relationship with the BRICS. Some are hopeful about future outcomes:

> As Chinese capitalism in Africa is poised to become a major story shaping the global economy, the spirit of Bandung, rather than the trite rhetoric of "colonialism", may provide a more relevant, productive and critical lens for observing development in the 21st century.
>
> (Kwan Lee 2009, 666)

However, extractive relations are largely being reinscribed by the BRICS powers.

During the commodity boom, Africa generally was composed of fast-growing but semi-democratic or semi-authoritarian (Young 2012) political economies – a model which some of the BRICS share. Part of what defines the BRICS is the size of their economies, with the partial exception of South Africa, although this is once again the largest economy in Africa, after the fall in oil prices and consequently the Nigerian currency. Obviously most African economies are unlikely to be able to rival the economic size or political power of the members of the BRICS grouping, at least in the medium term, again with perhaps the partial exception of Nigeria. The impact of the BRICS in Africa politically has been to strengthen extant regimes and open up more policy space for some neodevelopmental interventions. However, economically the impact of the BRICS has been to reinforce and deepen resource dependence. This absence of structural transformation means that the underlying structure of most African political economies remains intact. This in turn largely prevents possibilities for substantial poverty reduction. Consequently, while some African regimes are showing signs of movement towards more mixed economies, they do not appear to be *en route* to becoming brics. Even ones which are sometimes touted as developmental states, such as Botswana, have largely failed to diversify or transform their economies (Hillbom 2011). However, some argue that this is indeed happening through the vector of ICTs, and particularly mobile phones. Is this the case? Are these new ICTs fundamentally transforming the development context? This issue will be explored in the next chapter.

Note

1 Although this is now being rectified. See Taylor (2017), Stuenkel (2016), Stephen (2014), Golub (2016).

Further reading

Books

Moyo, D. 2009. *Dead Aid: Why Aid Is Not Working and How There Is Another Way for Africa*. London: Allen Lane.
Sachs, J. 2005. *The End of Poverty: How We Can Make It Happen in Our Lifetime*. London: Penguin.

Articles

Cammack, D. 2005. "The Logic of African Neopatrimonialism: What Role for Donors?" *Development Policy Review* 25 (5): 599–614.
Peiffer, C., and P. Englebert. 2012. "Extraversion, Vulnerability to Donors, and Political Liberalization in Africa." *African Affairs* 111 (444): 355–78.

Websites

Abdul Latif Jameel Poverty Action Lab, www.povertyactionlab.org
AIDDATA, www.aiddata.org

References

Action for Southern Africa, Christian Aid, and SCiAF. 2007. "Undermining Development? Copper mining in Zambia." www.actsa.org/Pictures/UpImages/pdf/Undermining%20development%20report.pdf.
Adem, S. 2010. "The Paradox of China's Policy in Africa." *African and Asian Studies* 9: 334–55.
Baird, S., C. McIntosh, and B. Ozler. 2009. *Designing Cost-Effective Cash Transfer Programs to Boost Schooling Among Young Women in Sub-Saharan Africa*. World Bank Policy Research Working Paper no. 5090. Washington, DC: World Bank.
Banerjee, A.V., and E. Duflo. 2012. *Poor Economics: Barefoot Hedge-Fund Managers, DIY Doctors and the Surprising Truth About Life on Less Than $1 a Day*. London: Penguin Books.
Barber, B. 1996. *Jihad vs. McWorld: Terrorism's Challenge to Democracy*. Ballantine Books.
Bauer, P.T. 1984. *Reality and Rhetoric: Studies in the Economics of Development*. Weidenfeld and Nicolson.
Biesheuvel, T. 2017. "This Miner's $190 Billion Tax Bill Would Take Centuries to Pay." www.bloomberg.com/news/articles/2017-07-24/acacia-gets-190-billion-tax-bill-it-would-take-centuries-to-pay.
Bodomo, A. 2009. "Africa-China Relations: Symmetry, Soft Power and South Africa." *The China Review* 9 (2): 169–78.
Bond, P., and K. Alam. 2010. "Grameen Bank and 'Microcredit': The 'Wonderful Story' That Never Happened." www.cadtm.org/spip.php?page=imprimer&id_article=6029.
Brautigam, D., and X.Y. Tang. 2011. "African Shenzhen: China's Special Economic Zones in Africa." *Journal of Modern African Studies* 49 (1): 27–54.

Carmody P. 2010. *Globalization in Africa: Recolonization or Renaissance?* Boulder, CO: Lynne Rienner Publishers.

———. 2013. *The Rise of the BRICS in Africa: The Geopolitics of South-South Relations.* London: Zed Books.

———, S. Mikhaylov, and N. Disandi. 2018. *Power Plays and Balancing Acts: The Paradoxical Effects of Chinese Trade on African Foreign Policy Positions.* Mimeo.

Corkin, L. 2013. *Uncovering African Agency: Angola's Management of China's Credit Lines.* Farnham and Burlington, VT: Ashgate.

Cowaloosur, H. 2014. "Land Grab in New Garb: Chinese Special Economic Zones in Africa." *African Identities.* doi:10.1080/14725843.2013.868674.

Easterly, W. 2006. *The White Man's Burden: Why the West's Efforts to Aid the Rest Have Done So Much Ill and So Little Good.* Oxford: Oxford University Press.

———. 2013. *The Tyranny of Experts: Economists, Dictators, and the Forgotten Rights of the Poor.* New York: Basic Books.

Esteban, M. 2010. "A Silent Invasion? African Views on the Growing Chinese Presence in Africa: The Case of Equatorial Guinea." *African and Asian Studies* 9: 232–51.

Fan Lim, K. 2010. "On China's Growing Geoeconomic Influence and the Evolution of Variagated Capitalism." *Geoforum* 41: 677–88.

Flores-Macías, G., and S. Kreps. 2013. "The Foreign Policy Consequences of Trade: China's Commercial Relations with African and Latin America, 1992–2006." *The Journal of Politics* 75 (2): 357–71.

French, H. 2014. *China's Second Continent: How a Million Migrants Are Building a New Empire in Africa.* New York: Knopf.

Glennie. J. 2008. *The Trouble with Aid.* London: Zed Books.

Godio, J. 2004. "The 'Argentine Anomaly': From Wealth Through Collapse to Neo-Developmentalism." *IPG* 2: 128–46.

Golub, P. 2016. *East Asia's Re-Emergence.* Cambridge: Polity.

Grimmett, R. 2010. *Conventional Arms Transfers to Developing Nations.* Washington, D.C.: United States Congressional Research Service. https://digital.library.unt.edu/ark:/67531/metadc93906/.

Guha, R. 2017. *India After Gandhi: The History of the World's Largest Democracy.* 2nd edition. London: Palgrave Macmillan.

Hart, G. 2014. *Rethinking the South African Crisis: Nationalism, Populism, Hegemony.* Durban: University of Kwa-Zulu Natal and University of Georgia Press.

Harte, J. 2012. "Turkey Shocks Africa, World Policy Institute." www.worldpolicy.org/journal/winter2012/turkey-shocks-africa.

Hart-Landsberg, M. 2013. *Capitalist Globalization: Consequences, Resistance and Alternatives.* New York: Monthly Review Press.

Harvey, D. 1982. *The Limits to Capital.* Oxford: Wiley-Blackwell.

Hillbom, E. 2011. "Botswana: A Development-Oriented Gate-Keeping State." *African Affairs* 111 (442): 67–89.

Jerven, M. 2013. *Poor Numbers: How We Are Misled by African Development Statistics and What to Do About It.* Cornell: Cornell University Press.

Kaplinsky, R. 2014. *Panel discussion on "Misconceptions, Realities and Unanswered Questions: China's Engagement with Africa."* London: Overseas Development Institute.

———., and D. Messner. 2008. "Introduction: The Impact of Asian Drivers on the Developing World." *World Development* 36 (2): 197–209.

Kennedy, P. 1989. *The Rise and Fall of Great Powers.* New York: Vintage Books.

Kohler, H., and R. Thornton. 2012. "Conditional Cash Transfers and the HIV/AIDS Prevention: Conditionally Promising?" *The World Bank Economic Review* 26 (2): 165–90.

Kolstad, I., and A. Wiig. 2011. "Better the Devil You Know? Chinese Foreign Direct Investment in Africa." *Journal of African Business* 12: 31–50.

Kragelund, P. 2014. "'Donors Go Home': Non-Traditional State Actors and the Creation of Development Space in Zambia." *Third World Quarterly* 35 (1): 145–62.

———, P. 2015. "Towards Convergence and Cooperation in the Global Development Finance Regime: Closing Africa's Policy Space?" *Cambridge Review of International Affairs* 28 (2): 246–262.

Kurlantzick, J. 2007. *Charm Offensive: How China's Soft Power Is Transforming the World*. New Haven and London: Yale University Press.

Kwan Lee, C. 2009. "Raw Encounters: Chinese Managers, African Workers and the Politics of Casualization in Africa's Chinese Enclaves." *The China Quarterly* 199: 647–66.

Magubane, B. 1996. *The Making of a Racist State: British Imperialism and the Union of South Africa 1875–1910*. Trenton, NJ: Africa World Press.

Mahajan, V. 2009. *Africa Rising: How 900 Million African Consumers Offer More Than You Think*. Upper Saddle River, NJ: Wharton School Publishing.

Masina, L. 2008. "Chinese Envoy's Remarks on Malawi Breed Resentment." *Voice of America*.

Mawdsley, E. 2012. *From Recipients to Donors: Emerging Powers and the Changing Development Landscape*. London and New York: Zed Books.

Morris, M., R. Kaplinksy, and D. Kaplan. 2012. *One Thing Leads to Another: Promoting Industrialisation by Making the Most of the Commodity Boom in Sub-Saharan Africa*. Self-published.

Moyo, D. 2009. *Dead Aid: Why Aid Is not Working and How There Is Another Way for Africa*. London: Allen Lane.

Nossiter, A. 2013. "China Finds Resistance to Oil Deals in Africa." *New York Times*.

Obiarah, N. 2007. "Who's Afraid of China in Africa?" In *African Perspectives on China in Africa*, eds. F. Manji and S. Marks. Oxford: Fahamu.

Ottaviani, J. 2013. "How Much African Land Is the UK Leasing?" *The Guardian*. www.theguardian.com/news/datablog/2013/nov/27/african-land-uk-investment.

Patey, L.A. 2014. *The New Kings of Crude: China, Indian and the Global Struggle for Oil in Sudan and South Sudan*. London: Hurst.

Power, M., G. Mohan, and M. Tan-Mullins. 2012. *China's Resource Diplomacy in Africa: Powering Development?* Basingstoke: Palgrave Macmillan.

Quinn, B. 2017. "UK Among Six Countries to Hit 0.7% UN Aid Spending Target." *The Guardian*. www.theguardian.com/global-development/2017/jan/04/uk-among-six-countries-hit-un-aid-spending-target-oecd.

Rowden, R. 2013. "Indian Agricultural Companies 'Land Grabbing' in Africa and Activists Response." In *Agricultural Develoment and Food Security in Africa: The Impact of Chinese, Indian and Brazilian Investments*, eds. F. Cheru and R. Modi. London: Zed Books.

Roy, A. 2010. *Poverty Capital: Microfinance and the Making of Development*. London: Routledge.

Rueschemeyer, D., E. Huber Stephens, and J.D. Stephens. 1991. *Capitalist Development and Democracy*. Polity.

Sachs, J. 2005. *The End of Poverty: How We Can Make It Happen in Our Lifetime*. London: Penguin Books.

———. 2015. *The Age of Sustainable Development*. New York: Columbia University Press.

Sharma, R. 2012. "Broken BRICs Why the Rest Stopped Rising." *Foreign Affairs* 91 (6): 2–7.

Southern Weekly. 2010. "Accusation of 'China Looting Resources' Groundless." an interview with Chinese ambassador to Zambia, 8 April 2010, translated and posted at the website of the Forum on China Africa Cooperation (FOCAC). http://www.focac.org/eng/jlydh/sjzs/t679463.htm, accessed 4 May 2010.

Stein, H. 2008. *Beyond the World Bank Agenda: An Institutional Approach to Development*. Chicago, IL: University of Chicago Press.

———. 2012. "Africa, Industrial Policy, and Export Processing Zones: Lessons from Asia." In *Good Growth and Governance in Africa: Rethinking Development Strategies*, eds. A. Noman, K. Botchwey, H. Stein and J. Stiglitz. Oxford: Oxford University Press.

———., S. Cunningham, and P. Carmody. 2018. *The Rise and Risks of the Random: The World Bank, Experimentation, and the African Development Agenda*. Mimeo.

Stephen, M. 2014. "Rising Powers, Global Capitalism and Liberal Global Governance: A Historical Materialist Account of the BRICs Challenge." *European Journal of International Relations* 20 (4): 912–38.

Stuenkel, O. 2016. *Post-Western World: How Emerging Powers Are Remaking Global Order*. Cambridge: Polity.

Taylor, I. 2014. *Africa Rising? BRICS – Diversifying Dependency*. Oxford: James Currey.

———. 2017. *Global Governance and Transnationalizing Capitalist Hegemony: The Myth of the "Emerging Powers"*. London and New York: Routledge.

Turkey, Republic of. 2014. www.hazine.gov.tr/default.aspx?nsw=EilDPQez15w=-H7deC+LxBI8=&mid=100&cid=16&nm=36.

van Rooyen, C., R. Stewart, and T. de Wet. 2012. "The Impact of Microfinance in Sub-Saharan Africa: A Systematic Review of the Evidence." *World Development* 40 (11): 2249–62.

Yanguas, P. 2018. *Why We Lie About Aid: Development and the Messy Politics of Change*. London: Zed Books.

Vehoeven, H. 2014. "Is Beijing's Non-Interference Policy History? How Africa Is Changing China." *The Washington Quarterly* 37 (2): 55–70.

Young, C. 2012. *The Postcolonial State in Africa: Fifty Years of Independence 1960–2010*. Madison: University of Wisconsin Press.

CHAPTER

8

ICT4D

Information technology for development?

> Far more people now have access to a mobile than have access to sewerage, piped water or electricity.
>
> (Mitullah *et al.* 2016 quoted in Heeks 2018, 1)

> It's modern-day slavery. It's smart slavery.
>
> (Uber driver, Cape Town, South Africa, June 2018)[1]

> Over the last 20 years, rather than reducing poverty, ICTs have actually increased inequality, and if 'development' is seen as being about the relative differences between people and communities, then it has had an overwhelming negative impact on development.
>
> (Unwin 2017, 1)

ICT4D

Perhaps no topic has generated as much interest and excitement in international development in recent decades as the so-called information and communication technology (ICT) "revolution". Technology can be defined as "devices or techniques that apply knowledge in order to complete a particular task", whereas "information is data that has been processed to make it useful to its recipient" (Heeks 2018, 7–9). The most widely diffused new ICTs are mobile phones, which represent a form of what is called "inverse infrastructure" because unlike other infrastructures, such as roads, fixed-line phone systems, piped sewage, etc., they travel with the user.

New, or relatively new, technologies like mobile phones and smart phones, which can connect to the Internet, are sometimes thought to fundamentally change the development context. For example, Jeffrey Sachs has argued that "mobile phones are the single most transformative technology for development" (quoted in Etzo and Collender 2010, 661). This may be correct, but in what ways are they transformative and of what? As Kransberg's first law states, technology is neither (inherently) good nor bad, nor is it neutral: it depends on the uses to which it is put. For example, new communication technologies may be used by Internet scammers (Burrell 2008) or by the Lord's Resistance Army in Central Africa to organise attacks (Crilly 2008).

ICT application in, and to, the field of international development is often called information technology for development (ICT4D). ICT4D implies that information technology can be used for international development. As Kleine (2013) notes, in the ICT4D literature ICTs are considered to be the means, whereas development is the end. For example, mobile phones may be used to deliver education or health interventions. However, information technology has much broader and wider impacts on international development than through these intentional types of interventions, on which there is an extensive and generally positive literature. For example, it affects people's livelihoods where raw materials for mobile phones, such as coltan, are sourced or where they are assembled. Likewise, how businesses use information technology affects their performance and thereby jobs, investment and other things which are important to economic and social development. Consequently, some people prefer to talk about information technology and international development (ICTD) rather than ICT4D. Is this really a revolution, though, in the sense of something which upends the previous socio-economic order, like the American or Russian revolutions? We will explore this question in this chapter.

ICT connection = development?

As noted earlier, the well-known geographer David Harvey notes that one of the defining features of globalisation is time–space compression. ICTs facilitate this by allowing people around the world, if they can afford it and have access it, to email people on the other side of the world and for the message to be received almost instantaneously. Such innovations have led certain writers to claim that "the world is flat" (Friedman 2005) or that when, combined with financialisation, this results in the "end of geography" (O'Brien 1992). While such claims are grossly exaggerated, as globalisation actually selectively intensifies uneven development, they have gained substantial credence in the ICT4D community. In some cases mobile phones are perceived to be a virtual "magic" or "silver bullet" to the problems of international development. For example, according to one author in a World Bank publication:

> The impact of these developments in ICT in Africa, in terms of both ICT development (increased infrastructure and access) and ICT for development (adoption of ICT applications), has been to advance the process of development itself, in terms of ICT for development. The result of this duality of sector transformation has been itself dually vast. On the one hand it has facilitated the delivery of services such as education, health, better governance (on the part of both leadership and governed), enterprise and business development, as well as their overall contribution to socioeconomic well-being (especially poverty reduction), political stability and self-actualization.
>
> (Opaku 2006, 153)

According to some authors, the Global South, or at least parts of it, have leapfrogged over fixed-line telecommunications, giving it a competitive advantage over the Global North. In 2005 less than a quarter of people in the Global South had mobile phone subscriptions, whereas by 2016 the figure had risen to 94 per cent (ITU 2016 cited in Heeks 2018). The figures for Internet penetration are less impressive. By way of example only 20 per cent or so of people in Tanzania have access, with 82 per cent of these doing so through their mobile phone (Ng'wanakilala 2018). The continental average for Africa is 35 per cent of the population with Internet access (Internet World Stats 2018).

While it is true that mobile phones, computers, email and the Internet have many vital applications in international development, such as in agriculture (see Duncombe 2018), their importance should not be overloaded or overstated. For example, the fact that people have mobile phones does not give them jobs or sources of livelihood, although they may be used to attain these, but their taking them may simply stop other people from getting them – another example of the fallacy of composition. Whether mobile phones result in improved livelihoods and welfare or may actually reduce these is dependent on the context in which they are adopted and social actors' positionality in relation to them. Connection into economic and social networks is not always necessarily a positive thing, as it may result in "adverse incorporation" (Bush 2007), where people's lives or livelihoods may actually disimprove. There are always winners and losers in technological development. For example, the widespread introduction of sewing machines in India put many tailors out of work, producing the appearance of "over-population" (Mamdani 1972).

James Ferguson (1990, 55) has an interesting take on why lack of connection to the global economy is often taken to be the central problematic of international development:

> The implicit argument is of the sort known to logicians as a fallacy of equivocation, of the form: (1) all banks have money; (2) every river has two banks; therefore, (3) all rivers have money. The fallacy of course consists in changing the meaning of one of the terms of the syllogism in the middle of the implication. The "development" version goes as follows: (1) poor countries are (by definition) "less developed"; (2) less developed countries are (by another definition) those which have not yet been fully brought into the modern economy: therefore, (3) poor countries are those which have not yet been fully brought into the modern economy.

However, one of the central problems of development in the Global South is not its lack of connection to the global economy, but the unfavourable and exploitative nature of its connections, with some estimates suggesting that over 93 per cent of Africa's exports are now in low-value-added primary commodities (UNCTAD n.d. in Stein 2014), a legacy of colonial underdevelopment.

Under colonialism, infrastructure in the Global South was oriented primarily toward meeting the needs of the metropolitan powers, such as the UK and France. This increased costs, as phone calls often had to be routed through Europe and back to neighbouring colonies in Africa if they were administered by a different foreign power, for example. In recent years, a number of fiber-optic cables have been rolled out around the continent. These have reduced overseas telecommunication costs substantially and enabled the development of new industries such as call centres in Ghana and South

Africa. The fact that there is a supply of English-speaking workers and the similarity in time zones with European countries makes these advantageous locations for call centre operations (Benner 2006). However, with the advent of the "fourth industrial revolution" involving robotisation, automation and artificial intelligence, some call centres in the Global South or their functions are being re-shored back to richer countries.

Much of the literature on ICT4D is boosterist in nature – that is, that they are seen to present a major opportunity for poverty reduction. Poverty is often seen to be the result of exclusion from the global economy, and once people are connected through their phones or the Internet, the problem is seen to be on its way towards resolution, if not actually resolved. However, this is very far from being the case. This is not to claim that the information "revolution" is having a similar impact to colonialism, but only to argue that connection by itself is not necessarily positive and that there is a need to think impacts through analytically and through evidence rather than making assumptions (Murphy and Carmody 2015).

Much of the Global South is experiencing thin integration ("thintegration") into the global informational economy (Carmody 2013; Murphy and Carmody 2015) by virtue of the fact that most of the research and development of ICTs takes place overseas in places such as Silicon Valley in the United States, and China assembles more than 90 per cent of the world's mobile phones and computers, according to some estimates. Many other countries such as Bangladesh or Myanmar then remain primarily as consumers of these technologies – evincing technological dependence and capital leakage, as natural resources are exchanged for foreign manufactured products, thereby replicating a colonial trade structure.

Foreign exchange is a scarce resource in many countries in the Global South and can be used to import machine tools to develop factories, rather than import consumer goods from overseas. During South Korea's industrialisation, exporting foreign currency illegally was punishable by death if the amount was over a million US dollars (Amsden 1989). Capital flows to foreign-owned service providers also represent a flow of social surplus from consumers to transnational corporations. Income is then flowing up the global (social) value chain from those in the informal sector in the Global South buying mobile phone credit to international stockholders, such as Sunil Mittal who now holds a majority stake in Bharti Airtel and is the fourteenth richest person in India, having previously been the sixth (Forbes 2011). Bharti Airtel is one of the biggest carriers in India, in addition to having operations in Africa and other countries in South Asia.

Impacts of mobile phones on livelihoods: a digital provide?

One of the assumptions made in many neoclassical economic models is that actors have perfect information – that is, that they complete knowledge about the price, quality and other features of the products they are buying, amongst other things. However, in reality this is not the case. For example, if you are buying a second-hand car you may not have information about how it runs. This is referred to as imperfect information

(Greenwald and Stiglitz 1986). Many markets are characterised by imperfect information, and mobile phones are often thought to help with this problem.

Probably the most famous paper about the impact of mobile phones on international development is one by Jensen (2007). In this paper he looked at the way fisherfolk in Kerala in South India were using their mobile phones to find out which beaches had the highest prices for fish before landing their catch. Thus, mobile phones were meant to benefit direct producers. He called this a digital provide. However, the paper has been the subject of stinging critique, with other researchers finding that fishermen did not use their mobile phones to ascertain market prices and that increased fish prices might have been the result of other factors, such as market trends or population growth (Steyn 2016). Another study found that larger boats chose where to land their catch, but that this was not the case for smaller ones (Srinivasan and Burrell 2015 cited in Heeks 2018).

One of the reasons often given for market failure in the developing world is asymmetric information between buyers and sellers (Lofren *et al.* 2002). Where markets are nonexistent or thin (poorly developed), competition and, consequently, access to information may be constrained. The use of market linkage and information systems (MLIS) is an attempt to circumvent this problem through the direct provision of information to farmers about prices, for example, or to connect producers to buyers to whom they would otherwise not have access.

MLISs are often thought to allow for disintermediation within value chains, cutting out middlemen, thereby allowing a greater share of the market price to be retained by the direct producers, reducing poverty. Prominent examples in an African context include Esoko in Ghana, which allows farmers to receive information about crop prices via text message, and ShopAfrica53, which has now closed down but accepted orders via the Internet and communicated this to (often) small producers via mobile phone text messages. In Chile, the government has attempted to ensure that small, medium, and micro-sized enterprises can access the government electronic procurement system Chilecompra – albeit with mixed success (Kleine 2013).

While such systems and businesses may reduce transaction costs and increase small-producer incomes, this may be done without necessarily expanding markets or resulting in the structural transformation of economies into higher, more diversified growth paths, as they have not primarily been used to increase productivity or produce better-quality products (Murphy and Carmody 2015). Heightened competition may benefit large producers that can take advantage of economies of scale. These services also represent a new form of intermediation or neo-intermediation, rather than disintermediation, and may create new forms of virtual or seemingly placeless power that may be more difficult to contest than those associated with other older forms of intermediation. In addition to "correcting" information assymetries, the other area which has probably received the most attention in the literature is money transfer services, which are meant to be another example of connection characterised by technological leapfrogging.

One study found that of the 20 countries worldwide where more than 10 per cent of the adult population had used "mobile money", 15 were in Africa (Godoy *et al.* 2012). In some countries, such as Gabon and Sudan, more than half the adult population reportedly uses mobile money. The iconic service is M-Pesa, operated by Safaricom in Kenya. This service allows customers to deposit, transfer and withdraw money from their mobile money accounts and was seeded by the United Kingdom's Department for International Development. There is a distinction between mobile money transfer systems and mobile banking, which offers a wider range of services to customers.

Rapid diffusion of mobile phones has led some to write about a "silicon savannah" in Kenya, partly also as a result of the extensive use of M-Pesa, through which almost half of the country's GDP transits during the course of a year (Munda 2017). Such statistics have led some to claim that Kenya, for example, has "leapfrogged" straight to the "digital age" (Ndemo and Weiss 2016). Such services have been slower to develop in richer country contexts because many people already have access to phone and Internet banking, for example. While such money transfer services may save time by obviating the need to take trips to bring money to relatives or reduce fees paid to traditional money transfer services, such as Western Union, they often do not fundamentally change the development context in most cases. However, in Somalia, one study found that 80 per cent of start-up capital from businesses came from the diaspora, and this was substantially remitted through mobile phones (Sheikh and Healy 2009).

A variety of other business and personal services are also being delivered by mobile phone, such as crop and health insurance. Mobile phone airtime is exchanged as a de facto form of money, particularly in countries where there are tighter restrictions on mobile banking. A service called ezetop (www.ezetop.com) enables mobile airtime to be purchased and transferred globally. This raises issues for macro-economic policy for developing countries since this represents, in effect, an alternative form of currency beyond state regulation. CurrencyFair (www.currencyfair.com) is an online system that allows for the purchase of foreign currency across borders.

There are many other ways in which mobile phones may affect livelihoods. For example, many jobs have been created around the world in mobile phone repair, unlocking, credit sales and other activities. However, these are low-productivity activities, often in the informal sector. Furthermore, the marginal productivity of labour, or the additional unit of output per unit of input, is very low in the informal sector, where the majority of African urban dwellers, for example, earn their living and ICTs do not change this.

Mo Ibrahim, who set up one of Africa's first mobile phone networks, Celtel, has said that he had hundreds of thousands of indirect employees. However, these were mostly informal-sector sellers of mobile phone credit scratch cards. Ironically Ibrahim also sponsors a yearly African governance index and awards a prize for the African leader who is adjudged to have governed best by virtue of criteria such as having stepped down voluntarily. This African leadership prize was meant to be awarded yearly when it was established in 2007, but has only been awarded five times at the time of this writing as a result of a shortage of suitable candidates. However, arguably by promoting the growth of the informal sector, mobile phone companies did not contribute to the development of a social contract where citizens pay taxes in return for government services (Leonard and Strauss 2003) – often thought to be the key to good governance. Uganda recently attempted to ban airtime scratch cards, but this was quickly reversed with Minister for Communications Frank Tumwebaze saying, "Let us not ban scratch cards abruptly. I agree that we should have walked along with our rural people" (quoted in itwebafrica.com). A controversial social media tax was introduced, according to President Museveni, to "stop gossip". Perhaps the regime is fearful of protests being organised via this medium. It was right to be worried, as extensive protests erupted in 2018 over the detention of popular opposition politician, Bobi Wine.

Whereas their manifold benefits are part of the reason for the rapid diffusion of mobile phones in the Global South, there are others as well. For example, casual day labourers may be increasingly recruited through calls to their mobile phones (Molony 2008). The costs of not being connected then may outweigh, the sometimes substantial costs of having a mobile phone, representing a form of "negative adoption" (Murphy and Carmody 2015). Data from rural South Africa show people spending more than a fifth of their total income on their mobile phones (Rey-Moreno *et al.* 2016 cited in Heeks 2018), whereas Diga's (2007) Ugandan study quotes a respondent who states that "mobile phones bring poverty". Students at the University of Dar-es-Salaam in Tanzania spend five times more on their mobile phones than they do on food (Kleine and Unwin 2009). However, what about more productive uses of the technology, such as e-business?

Development and e-business

There are also a number of definitions of what constitutes e-business. It is often thought to be synonymous with e-commerce provided by Internet-based stores or merchants. E-business, however, can be more broadly defined to include the multifold uses to which new ICTs are put by businesses in marketing or internal systems development and management. For example, many businesses in South Africa used to advertise their products on the social networking site Mxit, which at one point had seven times more daily posts than Twitter globally. According to some studies, approximately 90 per cent of e-commerce worldwide is conducted among businesses (B2B – business to business). However, with the rise of Facebook and Whatsapp in South Africa Mxit closed in 2015. Digital protectionism in China, with sites such as Facebook banned there, ensures a similar fate will not befall the WeChat network in that country, which has over a billion active monthly users. However, the established global advantages of companies like Facebook will make "digital industrialisation" in the Global South very difficult (Banga and Kozul-Wright 2018).

Some definitions of e-business or e-commerce are more expansive than others. Molla and Licker (2005, 90) define e-commerce as "conducting one or more core business functions internally within organizations or externally with suppliers, intermediaries, consumers, government, and other members of the enterprise environment through the application of solutions that run on Internet-based and other computer networks". Amit and Zott (2001) define an e-business as one that derives more than 10 per cent of its revenue from online sales. They argue that there are four value drivers for e-businesses: lock-in, novelty, complementarity and efficiency. Efficiency is enhanced through the reduction of transaction costs associated with conducting business online, whereas complementarity refers to the ability to market related products simultaneously. Lock-in is achieved when customers trust a website and consequently return to it, whereas novelty may be what initially captures the user's interest and attention.

E-business, broadly defined, has a diffuse but differentiated geography characterised by different levels of intensity of ICT usage. Whereas manufacturing activity has been extensively analysed using the concepts of global value and commodity chains and production networks, much e-business is concerned with trade in services, which is the largest economic sector in the so-called developed countries. Additionally, some have identified a quaternary sector that relates to the creation and sale of information or information-intensive services and goods. The geography of these activities cannot be captured adequately, however, through the idea of value chains. Some argue that value webs may be a more appropriate analytical frame for these types of economic activities (Ibach *et al.* 2005).

Certain types of e-business do relate to manufacturing activity. For example, in order to be productively integrated into global production networks, many firms may need systems such as electronic document interchange (EDI). These systems allow firms to share information about inventories that, for example, facilitate just-in-time delivery systems, thus reducing costs and raising efficiency. Vendor-managed inventory facilitated by EDI has been shown to reduce inventory or stock levels by as much as 40 to 45 per cent (Moodley 2002), thereby saving money. Short-cycle production and reduced lead times are increasingly central to competitive success in the global economy (Gibbon 2001). Thick inter-firm integration, which is ICT facilitated, is increasingly a prerequisite for participation in GPNs. Studies have found, however, that simply having access to a mobile phone may only marginally increase growth amongst micro-enterprises employing one or a few people (Chew *et al.* 2011). Robert Solow, a Nobel Prize–winning economist, once quipped that you could see the "computer age everywhere except in the productivity statistics". This is known as the productivity paradox, as people may spend time surfing the Internet rather than working in their offices, for example. One furniture manufacturer manager I interviewed in South Africa recounted a story of one of a company driver who sent 380 texts on a company phone to his girlfriend in a month.

A wider definition of e-business includes the development of electronic-based products in addition to illicit or illegal uses of ICTs, including mobile phones, to generate profit. Elements of e-business represent and facilitate a new mode of capital accumulation involving place-based processes in combination with "virtual" ones: "virtureal" accumulation. In an increasingly service-driven global economy where the distinction between tradable and non-tradable services is partially and unevenly breaking down as a result of this hybrid mode of capital accumulation, this has important implications for international development.

An example of this blurring is the purchase of hotel rooms through international websites. Tourism is a major growth focus for many developing countries. As China has emerged as the world's dominant manufacturing power, many other countries have found it difficult to compete in this sector. Economic activities that are complementary to, rather than competitive with, China's economic structure consequently have become increasingly important in many developing economies. Tourism is one of these and is one of the world's largest industries. It is, for example, India's largest foreign exchange earner. The fact that the "product" in tourism is the experience gives it certain features that make it particularly amenable to marketing and service delivery, such as booking tours via the Internet. Some refer to these developments as e-tourism.

Tourism services must still be consumed at the point of production but can be bought and traded internationally. In global tourism, a relatively small number of websites such as Expedia, Booking.com and Viator hold dominant market positions for booking flights, accommodation and tours. These websites often charge substantial commissions, and their dominance is explained both by brand reputation and tourist trust.

The fact that this virtual accumulation through a website is based on service delivery and experiences taking place in particular places (reality) has given rise to a new hybrid mode of capital accumulation that is virtureal. It often replicates previous patterns of economic extraversion, where economies are oriented to meet outsiders' needs, as foreign websites and travel companies often capture many of the gains that otherwise would accumulate locally in addition to those derived from advertising. Small and medium-sized touristic enterprises, however, often find it difficult not to list on foreign-owned websites, given their dominant market positions. Transnational web-based corporations increasingly capture value through this new mode of accumulation, leading to greater centralisation of capital. The prominent tour booking website company Viator, for example, is now owned by TripAdvisor, which is headquartered out of the US, where much of the profits flow. In fact, such developments have given rise to a new form of capital – virtual capital, explored in more detail later.

E-commerce – in a restricted definition, the trading of goods or services over the Internet – is relatively underdeveloped in the Global South. While certain industries, such as the Thai silk industry or the South African tourism industry, often have extensive web presences, these are somewhat exceptional. Perhaps more important are the ways in which new ICTs are being adopted, absorbed and adapted into previous business routines and practices in the developing world.

One useful way to think about the difference between ICT4D and e-business is provided by the distinction between imminent and immanent development explained earlier. ICT4D arguably refers to a form of imminent development where NGOs, governments and international agencies often work in collaboration with the private sector in their attempts to harness ICTs for developmental ends, whereas e-business can be understood to refer to a form of immanent development. There may, however, be elements of overlap and intersection between them – as in the case of the Dwesa project in the Eastern Cape Province of South Africa, where ICT4D project funds were used to train villagers in e-commerce skills (Day and Greenwood 2009). Common examples of ICT4D projects include government-sponsored and -supported telecenters in the developing world. In other examples, development agencies experiment with projects that enable aid recipients to receive money via mobile phones.

Case study: labour conditions in global factories

Conditions in "global factories" producing for the world market are often extremely exploitative, sometimes driving workers to have mass fainting events (Fuentes and Ehrenreich 1983) or committing suicide. This has been documented to be the case in many factories producing ICTs.

> The KYE factory in China produces manufactured goods for Microsoft and other U.S. factories. . . . Workers reported spending ninety-seven hours a week at the factory before the recession. . . . Workers race to meet the requirements of producing 2,000 Microsoft mice per shift. The factories are extremely crowded; one workshop, 105 feet by 105 feet, has almost 1,000 toiling workers. They are paid 65 cents an hour, with 52 cents an hour take-home pay, after the cost of abysmal factory food is deducted. Fourteen workers share each dorm room, sleeping on narrow bunk beds. They 'shower' by fetching hot water in a small plastic bucket for a sponge bath.
>
> (Foster and McChesney 2012, 64)

"Apple captures 53 per cent of the measured value from US sales and 47 per cent from sales outside in the US" (Dedrick *et al.* 2008, 20 quoted in Selwyn 2014, 126), but has also been found to rely on child labour in its supply chain (Garside 2013 cited in Selwyn 2014). In the case of its iPhone the profit margin was estimated to be an incredible 64 per cent on each one sold in the US (Xing and Detert 2010 cited in Hart-Landsberg 2013). Little wonder it has repeatedly been the most profitable company in the world. In the final quarter of 2017 Apple had revenue of US$88.3 billion (Heisler 2018); however, this is partly as a result of labour conditions in its supply chain.

Apple's Taiwanese-based supplier Foxconn, which assembles many of its products, was subject to much controversy recently. Apple executives praise the flexibility of the workforce in China, when thousands of people could be woken up in the middle of the night, given some tea and a biscuit and set to work replacing screens on iPhones (Hart-Landsberg 2013). Foxconn operates massive assembly operations on mainland China, with factories employing hundreds of thousands of workers. In 2010 there were 14 suicides at Foxconn plants as a result of low pay and long working hours. The company responded by putting nets under the upper floors of some of its factories. At times apparently workers had been pressurised to commit suicide by relatives who wanted to receive monetary compensation.

Apple responded to this scandal by taking some assembly work away from Foxconn, but its new supplier was reportedly paid even less and had worse wages and working conditions, raising Apple's profits (China Labour Watch 2015 cited in Selwyn 2017). The Chinese government also came under extreme pressure and was forced to allow independent trade union organising in the plants. Chinese workers' share of the final price of an iPhone is approximately 1.8 per cent (Kraemer, Linden and Dedrick 2011 cited in Selwyn 2017).

Mobile application (app) development

One of the areas which has received the most attention in the literature on ICTD has to do with mobile phone app development. Africa, at least since the era of colonialism, has been largely technologically dependent in the sense that research and development and innovations on the continent have been weak and technology has been mostly imported (Timamy 2007), although there have been some notable technological developments in recent years. For example, Anthony Mutua, a young Kenyan, developed a technology that charges mobile phones via specially designed shoes with microchips. This is an important innovation given the rapid spread of mobile technology and the irregularity and disjointed nature of electricity supply across much of the continent. There are also more than 170 emerging centres of technological development on the continent (Kelly and Firestone 2016), such as the i-hub in Kenya. Perhaps somewhat ironically, this Kenyan innovation hub grew out of violence since it was the not-for-profit company Ushahidi ("witness" in the East African language Swahili) that helped establish it. During the substantial post-election violence of 2007–2008 in Kenya, Ushahidi developed a website that allowed people to report violence via texts sent from their mobile phones. I-hub was founded by an American, Erik Hersman.

One of the applications that was developed at the i-hub is iCow, which allows farmers to access information via their mobile phones about where to find the nearest veterinarian or how best to tend to their cows (www.icow.co.ke). Other applications developed on the continent allow the "crowdsourcing" of other services, such as security. In Ghana, a text message to the service Hei Julor ("Hey thief") results in a security team being dispatched and up to ten friends or relatives being alerted to a burglary. Other innovative applications include M-Pedigree, which allows consumers to check if their medicines are real or counterfeit, which is important, as fake drugs kill tens of thousands of people in the developing world each year, and SimPill, which sends patients reminders via text to take their medicines at the appropriate times. While many of these services are important and convenient, they do not fundamentally change the structures of (under)development, and as noted earlier, new ICTs can also be used for illicit or illegal purposes, such as Internet scamming.

E-waste and Internet scamming

A variety of types of business are associated with the global informational economy that are either illegal or have substantial social and/or environmental costs. The world's electronic industry is the largest and fastest-growing industry, and according to the United States Environmental Protection Agency, over 100,000 computers are discarded daily in that country (cited in Oteng-Ababio 2012). A report by the United Nations Environment Programme found that up to 50 million tonnes of e-waste is generated annually, which would be enough to fill a line of dump trucks halfway around the globe (Grant and Oteng-Ababio 2012), and some estimates suggest that approximately 80 per cent of e-waste given to recyclers is exported to the developing world.

Sometimes attempts by governments to join the "global informational economy" have unintended effects, as when the government of Ghana zero-rated for tax purposes imports of secondhand computers, resulting in the country becoming a major destination for e-waste. Agbogbloshie, in the capital city Accra, is one of the primary sites for e-waste disposal and "reprocessing" and is highly polluted.

Computers contain precious gold and other metals that can be harvested and then used as a raw material for production of other commodities. The potentialities of this transformation of waste from cost to commodity have led some to argue that it may be possible to establish industries around the reprocessing of e-waste, as in South Africa. Some of the competitive advantages which the developing world offers for this trade, however, are comparative labour and environmental underregulation which reduce costs. Formalisation of an industry may reduce these "advantages" and result in the redirection of the trade.

There are also forms of illicit e-business such as the infamous 419 scams, named after the relevant section of the Nigerian criminal code. These are attempts to lure via emails unsuspecting or greedy users into parting with money in return for access to larger sums in the future, after perhaps being misappropriated by corrupt officials or their relatives. While the impersonation of officials and fraud is illegal, some argue that there is a moral economy to 419 scams since they can be seen as attempts to extract resources from the West in response to the "looting" of Africa by Westerners in collaboration with corrupt local government officials (Glickman 2005).

Other forms of Internet scamming include those where young men impersonate a woman in the hope of securing a foreign "boyfriend" for financial gain. In some instances, these men will type responses to the mark/victim while a woman is positioned in front of a webcam (Burrell 2008). Thus, the Internet both disembodies and re-embodies social relations. This form of "business" entails the trade of emotional gratification for money – a virtual market for love – even if that is not the understanding of all participants. ICTs then may be used to both reduce and increase inequalities, with the World Trade Organization noting that a lack of digital and technological capabilities may cement and widen the technology divide (WTO 2017 cited in Banga and Kozul-Wright 2018).

Gender and e-business

Twenty-five per cent less women than men have Internet access in Africa, representing a gender digital divide (ITU 2017). The ability to take advantage of and develop electronically enabled or electronic-based business is embedded in, and constrained by, social structures such as gender. Access to capital, gender norms, literacy and other issues may constrain women's ability to develop e-businesses. In Cameroon in West Africa in the early 2000s, for example, it was found that the majority of women entrepreneurs were unaware of the Internet (Yitaben and Tchinda 2009). The International Telecommunication Union has produced statistics on the gender digital divide showing substantial differential usage of ICT by gender, and new digital divides are emerging shaped by social structures other than whether a country is considered "developed" or not. The extent of the gender digital divide also varies between and within regions. For example, in relation to Internet access the digital gender divide in the six countries studied by Alozie and Akpan-Obong (2017) varied from 3.4 per cent in Uganda to 21 per cent in Ghana.

Chew *et al.* (2011) found in their sample in India that only 10 per cent of female micro-entrepreneurs regularly used their mobile phone for business purposes and that this had a marginal relationship and impact on firm growth – limited use resulted in limited impact. There are two potential explanations for this. The first is that there is an information deficit and that these female entrepreneurs were not aware of the potential benefits of mobile phone usage for their businesses. The second is that mobile phones do not enable substantial business gains and, consequently, are not frequently used. These explanations are not, however, mutually exclusive.

The virtual economy

The virtual economy proper is a distinctive form of e-business in that it takes place entirely online. An example is where gamers, often in rich countries, pay people in developing countries to develop their characters (power levelling) in online games such as *World of Warcraft*. Payment for this service may be through virtual money in the game. There are also other virtual currencies such as Bitcoin which can be used to purchase products and services over the Internet. Concerns have been expressed about these currencies because they largely escape state regulation; the Thai government, for instance, declared Bitcoin illegal. The government was reportedly concerned that allowing the circulation of alternative currencies could affect the exchange rate of the Thai bhat and, consequently, the country's development.

The rise of the 'sharing economy': virtual capital

Microwork is an emerging mode of organisation whereby large tasks are passed out to multiple employees and their output (re)assembled digitally. The term "microwork" was developed by the founder of Samasource, a non-profit microwork company. An example of microwork might be harvesting email addresses from the Internet and then compiling them into a database. Concerns have been raised, however, that private, for-profit companies may circumvent labour laws and minimum wages through the use of piece rates in microwork as a result of being able to source cheap labour globally (LeVine 2013). Samasource says on its website that it is committed to paying a living wage. However, this is not the case with other platforms, and in some cases workers work for free in order to try to get good ratings on the system so they can get paying work in the future (Graham *et al.* 2017). Digital technologies are changing the nature of work in other ways as well.

Another emergent trend in the international development literature is a focus on the so-called "sharing economy". Uber is an American-based company which has been instrumental in changing, if not revolutionising, mobility in many cities around the world. It has, however, been banned in certain cities and countries around the world over concern about some of its practices and that it presents "unfair competition" to taxi drivers (Rhodes 2017). It is banned in Namibia because of "concerns over foreign profiteering and tax avoidance" (Moskvitch 2018). In discussion with a taxi driver in London, it was noted that they must have licences from the local authority to operate and have comprehensive insurance, thereby raising their costs in comparison. However, there was a recent court ruling in the UK that Uber is an employer and that consequently its drivers are entitled to the national living wage.

Uber is part of the so-called sharing economy: marketing itself as a technology, rather than a taxi company, and has recently extended its operations to several countries in Africa, in addition to operating in other parts of the globe. There are also a number of other "ride-sharing" apps such as Taxify which operate globally. Taxify, which is headquartered in Estonia, also now has operations in Kenya and South Africa and an "activation hub" in Nigeria. These companies are implicated in the rise of technologically mediated urbanism across the world.

According to Graham *et al.* (2017, 140):

> a key feature of digital work platforms is that they attempt to minimise the outside regulation of the relationship between employer and employee. . . . These issues are particularly acute when transactions cross national borders: as it becomes unclear which jurisdictions' regulations apply to the work being transacted.

In this way new information and communication technologies are facilitating further deregulation of the business–labour relationship. In June 2018 a survey was conducted of 26 Uber and Taxify drivers in Cape Town, South Africa. Twenty-five of the twenty-six drivers interviewed were male and one was female (3.8 per cent of the sample). This may relate partly to perceptions around gender and work, as taxi driving has traditionally been male dominated around the world. However, there are also initiatives to disrupt this this gender imbalance in professional driving. For example, Little Cabs in Nairobi, which is an online ride-sharing platform, offers customers the choice of a male or female driver and has seen a 13-fold increase in the number of women drivers in the past two years (Bhalla 2018). It is estimated that women account for only 3 per cent of the city's "e-taxi" drivers.

Graham and Anwar (2018) note many of the drivers they spoke to experience hardship and have to work long hours, with some sleeping in their cars over the weekend so that they could maximise the amount of time they could spend driving customers home. This result was also reflected in our survey, where the average number of hours driving per week was 64, equating to approximately 13 hours a day over a five-day week or almost 11 hours a day over a six-day week. Some of the drivers were working more than 70 hours a week, representing a very high level of "self-exploitation". One of the drivers quoted at the beginning of the chapter equated it to slavery. Whereas the "sharing economy", as noted earlier, is meant to have an inclusionary ethos, the reality is much different, characterised by very highly exploitative labour conditions, which are technologically mediated and defined by high insecurity for many drivers in the sector.

In his famous book *The Mystery of Capital*, de Soto (2000) argued that many of the assets of the poor constituted 'dead capital', which could be unlocked by them being given property rights to their informal housing, which they could then use as collateral to start small businesses, for example. In a sense the sharing economy is unlocking forms of "dead" or creating new types of capital. However, much of the value is being unlocked by virtual capital, which will be elaborated on later. This virtual capital arises not just as a result of "social innovations" facilitated by new ICTs but also broader patterns of flexible accumulation. We can see these ideas around individualism and flexibility flowing into the rhetoric around the "sharing economy", such as AirBnB, even though some of these practices can be highly exploitative and based on digital sweatshops, as noted earlier (Ettlinger 2014, 2016). Ettlinger (2017, 61) in fact goes so far as to argue that these innovations are "part of an emergent regime of accumulation, overlaying and co-existing with flexible production". However, she does not name what this emergent regime is. One way to think of it is through the idea of virtual capitalism.

Virtual capital is capital which extracts value (or economic rent) through technological intermediation, which allows control over the capital (in the conventional sense of income-bearing assets) and labour of others. As such it is extractivist. Whereas in conventional business models often three main actors are involved in production: capital, labour and management, virtual capitalist business models largely do away with management as a social actor. Rather, management is undertaken by a combination of technology and customers, who rate the services they receive via their mobile phone apps. This then is often a way of extracting value from petty service producers, who may combine ownership of some capital, such as a car, with self-exploitation (having to drive for over 60 hours a week on average in our survey). Where other individuals or companies own the cars, rather than the drivers themselves, this puts further pressure on returns to labour and living standards.

In this model profits flow off-shore to companies such as Uber or Taxify, who because they are not in a direct employment relation with their drivers, are difficult to contest with. The drivers in our survey in Cape Town noted that Uber did not even provide a phone number if you were having difficulty and you had to contact them online. Some drivers can put different services in competition with each other in order to try to retain more of the value which they generate, and this may increase as more ride-sharing applications become available or widely used. However, at the moment there is a pronounced power imbalance.

Transformative technologies?

Often ICT4D projects have failed to fulfil expectations for the reasons detailed by Heeks (2006) and Kleine (2013). Often ICT4D projects fail because of lack of demand by the intended target group and the concomitant excessive focus on supply-side measures, with a consequent lack of financial sustainability over the longer term. Common examples of these would include Internet or telecentres.

Mobile phones and other "new" ICTs are not a magic bullet for development. They are inserted into pre-existing social structures, patterns and processes. Most of the economic value which arises from the production and use of these new technologies is captured by major TNCs. People getting mobile phones does not create jobs for them. They may help certain small businesses, but again the fallacy of composition comes into play: mobile phone–enabled businesses may simply take customers from other businesses without growing the overall size of the economy or "pie". Productivity growth and the wide distribution of the fruits of that are what create more equitable or distributed forms of development. While mobile phones do allow some time savings for certain people – for example, not having to take days to travel with remittances to get them to relatives in rural areas – the overall impacts of this on the economy are somewhat limited as they don't foster the creation of new better or higher-quality products. While there have been and continue to be many innovative applications of mobile phones in development, they are not the transformative technologies that some claim they are. Rather, they often reinforce existing patterns of dependence and social inequality. For example, importing mobile phones and associated infrastructure such as base transceiver stations into Africa represents a drain of scarce foreign currency and has to be paid for largely by exports of raw materials. This then represents a reinscription or reinforcement of colonial economies – importing manufactured goods and exporting raw materials in order to be able to pay for them. Mobile phones may be socially transformative by allowing easier social interaction at a distance. They and other new ICTs are also economically transformative in those world regions where they are produced, developed and designed. However, this is not the case for those world regions which are primarily integrated into ICT GPNs as consumers, despite the hype often associated with them.

As noted earlier, technology is resources transformed. Intuitively a high natural resource endowment should be a developmental boon. "Gifts of nature", such as oil or gold deposits, should provide foreign exchange, income, jobs and other benefits to the regions in which they are found. However, this is not always necessarily the case. Why not? We now turn to explore this issue in the next chapter.

Note

1 The survey/interviews this quote is drawn from was conducted by Alicia Fortuin at the African Centre for Cities at the University of Cape Town and funded by the University of Johannesburg.

Further reading

Books

Heeks, R. 2018. *Information and Communication for Development (ICT4D).* London: Routledge.

Unwin, T., ed. 2009. *ICT4D: Information and Communication Technology for Development.* Cambridge: Cambridge University Press.

Articles

Aker, J., and I. Mbiti. 2010. "Mobile Phones and Economic Development in Africa." *Journal of Economic Perspectives* 24 (3): 207–32.

Jensen, R. 2007. "The Digital Provide: Information (Technology), Market Performance, and Welfare in the South Indian Fisheries Sector." *Quarterly Journal of Economics* 122 (3): 879–924.

Websites

Banga, Rashmi, and Richard Kozul-Wright. 2018. *South-South Digital Cooperation for Industrialization: A Regional Integration Agenda.* UNCTAD/GDS/ECIDC/2018/1. Geneva: United Nations Conference on Trade and Development. https://unctad.org/en/PublicationsLibrary/gdsecidc2018d1_en.pdf

Digital Development, World Bank, www.worldbank.org/en/topic/digitaldevelopment

Information Technologies and International Development, https://itidjournal.org/index.php/itid

References

Alozie, N.O., and P. Akpan-Obong. 2017. "The Digital Gender Divide: Confronting Obstacles to Women's Development in Africa." *Development Policy Review* 35 (2): 137–60.

Amit, R., and C. Zott. 2001. "Value Creation in E-Business." *Strategic Management Journal* 22: 493–520.

Amsden, A. 1989. *Asia's Next Giant: South Korea and Late Industrialization.* New York and Oxford: Oxford University Press.

Banga, Rashmi, and Richard Kozul-Wright. 2018. *South-South Digital Cooperation for Industrialization: A Regional Integration Agenda.* UNCTAD/GDS/ECIDC/2018/1. Geneva: United Nations Conference on Trade and Development.

Benner, C. 2006. "South Africa on-Call: Information Technology and Labour Restructuring in South African Call Centres." *Regional Studies* 40 (2): 1025–40.

Bhalla, N. 2018. *Women Cabbies Hit Nairobi's Roads as Taxi-Hailing Apps Mushroom.* www.reuters.com/article/us-kenya-women-taxi-drivers/women-cabbies-hit-nairobis-roads-as-taxi-hailing-apps-mushroom-idUSKBN1I300D.

Burrell, J. 2008. "Problematic Empowerment: West African Internet Scams as Strategic Misrepresentation." *Information Technologies & International Development* 4 (4): 15–30.

Bush, R. 2007. *Poverty and Neoliberalism: Persistence and Reproduction in the Global South.* London: Pluto.

Carmody, P. 2013. "A Knowledge Economy or an Information Society in Africa? Thintegration and the Mobile Phone Revolution." *Information Technology for Development* 19 (1): 24–39.

Chew, H., M. Levy, and V. Ilavarasan. 2011. "The Limited Impact of ICTs on Microenterprise Growth: A Study of Businesses Owned by Women in Urban India." *Information Technologies and International Development* 7 (4): 1–16.

Crilly, R. 2008. "People of Congo Suffer in Someone Else's War." *Irish Times*, October 18th.

Day, B., and P. Greenwood. 2009. "Information and Communication Technologies for Rural Development." In *ICT4D: Information and Communication Technology for Development*, ed. P.T.H. Unwin. Cambridge, UK: Cambridge University Press.

Dedrick, J., K. Kraemer, and G. Linden. 2008. "Who Profits from Innovation in Global Value Chains? A Study of the iPod and Notebook PCs." In *Sloan Industry Studies Annual Conference.* Boston, MA.

Diga, K. 2007. *Mobile Phones and Poverty Reduction: IDRC Field Study.* www.slideshare.net/kdiga/mobile-cell-phone-poverty-reduction-in-africa (last accessed 22 July 2013).

Duncombe, R., (ed.) 2018. *Digital Technologies for Agricultural and Rural Development in the Global South.* Croydon: CABI.

Ettlinger, N. 2014. "The Openness Paradigm." *New Left Review* (89): 89–100.

———. 2016. "The Governance of Crowdsourcing: Rationalities of the New Exploitation." *Environment and Planning A* 48 (11): 2162–80.

———. 2017. "Open Innovation and Its Discontents." *Geoforum* 80: 61–71.

Etzo, S., and C. Collender. 2010. "The Mobile Phone 'Revolution' in Africa: Rhetoric or Reality?" *African Affairs* 109 (437): 659–68.

Ferguson, J. 1990. *The Anti-Politics Machine: "Development," Depoliticization, and Bureaucratic Power in Lesotho.* Cambridge: Cambridge University Press.

Forbes. 2011. "India's Richest." October 26th. www.forbes.com/lists/2011/77/india-billionaires-11_Sunil-Mittal_EM57.html (last accessed 1 January 2014).

Foster, J.B., and R. McChesney. 2012. "The Global Stagnation and China." *Monthly Review-an Independent Socialist Magazine* 63 (10): 61–4.

Friedman, M. 2005. "The Methodology of Positive Economics." In *Philosophy of Economics: An Anthology*, ed. D. Hausman. Cambridge and New York: Cambridge University Press, 3rd edition, pp. 145–78.

Fuentes, A., and B. Ehrenreich. 1983. *Women in the Global Factory*. Boston: South End Press.

Garside, J. 2013. "Child Labour Uncovered in China's Supply Chain." *The Guardian*. www.guardian.co.uk/technology/2013/jan/25/apple-child-labour-supply.

Gibbon, P. 2001. *At the Cutting Edge? UK Clothing Retailers' Global Sourcing Patterns and Practices and Their Implications for Developing Countries*. Copenhagen, Denmark: Centre for Development Research, Mimeo.

Glickman, H. 2005. "The Nigerian '419' Advance Fee Scams: Prank or Peril?" *Canadian Journal of African Studies* 39 (3): 460–89.

Godoy, J., B. Tortora, J. Sonnenschein, and J. Kendal. 2012. "Payments and Money Transfer Behavior of Sub-Saharan Africans." www.microfinancegateway.org/sites/default/files/mfg-en-paper-payments-and-money-transfer-behavior-of-sub-saharan-africans-jun-2012.pdf: Gallup and Bill and Melinda Gates Foundation.

Graham, M., and M. Anwar. 2018. "Towards a Fairer Sharing Economy." In *The Cambridge Handbook of Law and Regulation of the Sharing Economy*, eds. N. Davidson, M. Finck, and J. Infranca. Cambridge: Cambridge University Press.

———, I. Hjorth, and V. Lehdonvirta. 2017. "Digital Labour and Development: Impacts of Global Digital Labour Platforms and the Gig Economy on Worker Livelihoods." *Transfer-European Review of Labour and Research* 23 (2): 135–62.

Grant, R., and M. Oteng-Ababio. 2012. "Mapping the Invisible and Real 'African' Economy: Urban E-Waste Circuitry." *Urban Geography* 33 (1): 1–21.

Greenwald, B., and J. Stiglitz. 1986. "Externalities in Economies with Imperfect Information and Incomplete Markets." *Quarterly Journal of Economics* 90: 229–64.

Hart-Landsberg, M. 2013. *Capitalist Globalization: Consequences, Resistance and Alternatives*. New York: Monthly Review Press.

Heeks, R. 2006. "Theorizing ICT4D Research." *Information Technologies and International Development* 3 (3): 1–4.

———. 2018. *Information and Communication for Development (ICT4D)*. London: Routledge.

Heisler, Y. 2018. "Apple Posts Monster Earnings with $88.3 Billion in Revenue; iPhone Sales Fall Short." https://bgr.com/2018/02/01/apple-earnings-q4-2017-iphone-sales-revenue/.

Ibach, P., M. Horbank, and G. Tamm. 2005. "Dynamic Value Webs in Mobile Environments Using Adaptive Location-Based Services." *Thirty-Eigth Annual Hawaii International Conference on System Sciences (HICSS-38)*. Value Webs in the Digital Economy Mini-Track, Hawaii, January.

Internet World Stats. 2018. "Internet Penetration in Africa, 31st December 2017." www.internetworldstats.com/stats1.htm.

ITU. 2017. *ITU Facts and Figures 2017*. www.itu.int/en/ITU-D/Statistics/Documents/facts/ICTFactsFigures2017.pdf.

Jensen, R. 2007. "The Digital Provide: Information (Technology), Market Performance, and Welfare in the South Indian Fisheries Sector." *Quarterly Journal of Economics* 122 (3): 879–924.

Kelly, T., and R. Firestone. 2016. *How Tech Hubs Are Helping to Drive Economic Growth in Africa*. WDR 2016 Background Paper: World Bank. Washington, DC: World Bank. https://openknowledge.worldbank.org/handle/10986/23645. License: CC BY 3.0 IGO.

Kleine, D. 2013. *Technologies of Choice? Icts, Development, and The Capabilities Approach*. Cambridge, MA: MIT Press.

———., and T. Unwin. 2009. "Technological Revolution, Evolution and New Dependencies: What's New About ICT4D?" *Third World Quarterly* 30 (5): 1045–67.

Leonard, D.K., and S. Strauss. 2003. *Africa's Stalled Development: International Causes and Cures*. Boulder, CO: Lynne Rienner Publishers.

LeVine, M. 2013. *In Palestine, "Death" by a Thousand Micro Jobs*. www.aljazeera.com/indepth/opinion/2013/04/201348101128647355.html.

Lofren, K.G., J. Persson, and J. Weibull. 2002. "Markets with Asymmetric Information: The Contributions of George Akerlof, Michael Spence and Joseph Stiglitz." *The Scandinavian Journal of Economics* 104 (2): 195–211.

Mamdani, M. 1972. *The Myth of Population Control. Family, Caste, and Class in an Indian Village*. New York and London: Monthly Review Press.

Molla, A., and P. Licker. 2005. "Maturation Stage of E-Commerce in Developing Countries: A Survey of South African Companies." *Information Technologies and International Development* 2 (1): 89–98.

Molony, T. 2008. "The Role of Mobile Phones in Tanzania's Informal Construction Sector: The Case of Dar es Salaam." *Urban Forum* 19 (2): 175–86.

Moodley, S. 2002. "E-Business in the South African Apparel Sector: A Utopian Vision of Efficiency?" *The Developing Economies* 40 (1): 67–100.

Moskvitch, K. 2018. "Volkswagen's Got a Radical Plan to Fix Ride-Sharing and Car Ownership." *Wired*. www.wired.co.uk/article/volkswagen-car-sharing-rwanda-africa.

Munda, C. 2017. "Transactions Through Mobile Money Platforms Close to Half GDP." www.nation.co.ke/business/Yearly-mobile-money-deals-close-GDP/996-4041666-dtaks6z/index.html.

Murphy, J.T., and P. Carmody. 2015. *Africa's Information Revolution: Technical Regimes and Production Networks in South Africa and Tanzania*. Chichester, West Sussex and Malden, MA: John Wiley & Sons Inc.

Ndemo, B., and T. Weiss. 2016. *Digital Kenya: An Entrepreneurial Revolution in the Making*. Basingstoke and New York: Palgrave Macmillan.

Ng'wanakilala, F. 2018. *Tanzania Internet Users Hit 23 million 82 Percent go Online Via Phones Regulator*. www.reuters.com/article/us-tanzania-telecoms/tanzania-internet-users-hit-23-million-82-percent-go-online-via-phones-regulator-idUSKCN1G715F.

O'Brien, R. 1992. *Global Financial Integration: The End of Geography*. London: [Published for] Royal Institute of International Affairs [by] Pinter.

Opaku, J. 2006. "Leapfrogging into the Information Economy: Harnessing Information and Communications Technologies in Botswana, Mauritania, and Tanzania." In *Attacking Africa's Poverty: Experience from the Ground*, eds. M.L. Fox and R. Liebenthal. Washington, DC: World Bank.

Oteng-Ababio, M. 2012. "Electronic Waste Management in Ghana – Issues and Practices." In *Sustainable Development: Authoritative and Leading Edge Content for Environmental Management*, ed. S. Curkovic. www.intechopen.com/books/sustainable-development-authoritative-and-leading-edge-content-for-environmental-management/electronic-waste-management-in-ghana-issues-and-practices.

Rhodes, A. 2017. "Uber: Which Countries Have Banned the Controversial Taxi App." *The Independent*. www.independent.co.uk/travel/news-and-advice/uber-ban-countries-where-world-taxi-app-europe-taxi-us-states-china-asia-legal-a7707436.html.

Selwyn, B. 2014. *The Global Develoment Crisis*. Cambridge: Polity.

———. 2017. *The Struggle for Development*. Cambridge and Malden, MA: Polity.

Sheikh, H., and S. Healy. 2009. "Somalia's Missing Million: The Somali Diaspora and Its Role in Development." www.undp.org/content/dam/somalia/docs/undp_report_onsomali_diaspora.pdf: UNDP.

Steyn, J. 2016. "A Critique of the Claims About Mobile Phones and Kerala Fisherman: The Importance of the Context of Complex Social Systems." *The Electronic Journal of Information Systems in Developing Countries* 74 (3): 1–31.

Timamy, M.H.K. 2007. *The Political Economy of Technological Underdevelopment in South Africa: Renaissance Prospects, Global Tyranny, and Organized Spoilation*. Lagos: CBAAC.

Unwin, T. 2017. *Reclaiming Information and Communication Technologies for Development*. Oxford and New York: Oxford University Press.

Yitaben, G., and E. Tchinda. 2009. "Internet Use Among Women Entrepreneurs in the Textile Sector in Douala, Cameroon: Self-Taught and Independent." In *African Women and ICTs: Investigating Technology, Gender and Empowerment*, eds. I. Buskens and A. Webb. London: Zed Book; Pretoria: Inisa Press.

CHAPTER

9

"The resource curse"

Land, wealth and politics

One of the concepts which has gained the most traction in development studies in recent decades is the so-called resource curse. Whereas intuitively substantial natural resource endowments should be a boon to development, they are often associated with deepening poverty, authoritarianism, conflict and lower economic growth. In an African context this is sometimes referred to as the "paradox of plenty" (Karl 1997). The term "resource curse" was coined by the economic geographer, Richard Auty (1993).

There are a variety of channels through which the resource curse is meant to operate. These can be broken down into economic, political, social and environmental, although these are really what social scientists call heuristics, or ways of approximating causal dynamic or breaking-up analysis, as there is no neat compartmentalisation between different spheres. This means more critical scholars prefer to talk of political economy. As noted earlier, some scholars now refer to "socio-nature" rather than society or nature, as it can be argued that this is an artificial division and that what constitutes nature is socially constructed and defined and that taking humans out of nature allows or encourages us to exploit resources unsustainably. "Deep ecologists", such as Næss (1989), argue that humans have no inherent right to subordinate or dominate nature. There are, however, counter-arguments.

The well-known Oxford economist, Paul Collier (2011), argues that if revenues from resources can be invested to deliver improved living standards and "permanent income", then it is worth it. In a talk I went to he said by way of example that it was worth clearing a forest to build Oxford University where he works.

Economic channels of the resource curse

Economically, one of the main mechanisms through which the resource curse is meant to operate is through the so-called "Dutch disease", named after the process that took place after the Netherlands discovered substantial gas deposits in 1959. The way in which it works is quite simple. Given the fact that the US is the world's largest economy (at least when measured in US dollars; China is when measured at purchasing power parity, or in terms of the value of what local currency can buy there);consequently the US dollar is the world's most important reserve currency, as most commodities around the world, such as oil are denominated in dollars. As much international trade is denominated in dollars, for oil or copper, for example, countries need to hold reserves of dollars – and hence reserve currency. If a currency is free-floating (i.e. its value is not regulated by the government), its exchange rate with other currencies will be determined by the law of supply and demand. If the US government prints more dollars, the value will go down, for example, all other things being equal. On the other hand, if the US central bank raises interest rates, (some) overseas investors will want to buy more dollars to invest in banks there to garner higher interest rates, again all other things being equal.

When there is a natural resource windfall, such as a major oil discovery, this affects the currency exchange rate as US dollars come into the economy in substantial amounts to pay for the oil or other resource which is being exported. This makes the US dollar relatively more abundant in the economy, so its value decreases in relative terms. Another way to put this is that the value of the local currency goes up, or appreciates. What are the impacts of this? Let us take the example of the Netherlands.

When gas was discovered and began to be exploited, the value of the Dutch gilder went up. This made Dutch-produced televisions more expensive overseas, while making it cheaper for residents of the Netherlands to buy German TVs, for example. Thus, much of the manufacturing industry in Holland was subject to intense competitive pressure from overseas producers and some firms failed as a result. One typical impact of the resource curse is the narrowing of the economy as the natural resource sector booms and manufacturing is undercut. This impact is accentuated by another economic dimension of the resource curse – the so-called "resource pull" effect. The booming resource sector may also attract capital and labour away from other sectors of the economy, thereby undercutting them.

There are other economic aspects of the resource which receive less attention. Natural resource extraction is often capital, rather than labour, intensive, although that is not the image we often have of gold mining, for example. The Kizomba, a floating, production, storage and offloading platform (FPSO) or vessel used off the cost of Angola, by itself cost nearly a billion US dollars (Figure 9.1). On land, large-scale mining also requires major capital outlays and is becoming more capital intensive. For example, some mines in South Africa are now introducing robots to replace humans, given the dangers associated with going deeper to extract largely worked-out mines. As an article about South Africa explains:

> Soon, perhaps, machines will replace the nearly half a million men who toil daily in conditions where the very walls conspire to kill them. As mines go deeper so the risk of a "rock burst" increases – an explosive fracturing of rock that will crush a man in a second. . . . Death comes in many forms underground; drowning because of flooding, being struck by moving vehicles and even fire. In 2014 eight miners died from smoke inhalation after an electrical fire broke out at Harmony Gold's Doornkop gold mine, west of Johannesburg.
>
> (du Venage 2017)

FIGURE 9.1 Kizomba: a floating, production, storage and offloading platform

Given the risks associated with mining, miners often discount the risks associated with HIV (Campbell 2003). As they often work in Southern Africa, for example, in very oppressive, racialised and dangerous conditions, they may seek to exert power over others, such as sex workers through practices such as "flesh on flesh" sex, which raises the possibility of HIV transmission. This in turn has economic costs for the South African state through increased health care expenditures. There is also what economists call an opportunity cost associated with this, as governments have to spend money on health care instead of things like economic infrastructure, which might boost economic growth, in addition to the devastating social impacts and associated intense human suffering.

Resource processing also tends to be highly capital intensive. For example, MOZAL, the major aluminium processing plant in Mozambique, only employed 1,910 workers in 2009 but accounted for fully 40 per cent of the country's exports (Nino and Le Billon 2014). The plant cost US$2 billion to build. This plant has the characteristics of what is often called an enclave – that is a relatively small economic area which sources few inputs from, and has few forward linkages to, the local economy. Often in these type of enclaved economies skilled foreign workers are imported and relatively few locals are employed. Likewise, much of the technology used is imported from overseas, with the local economy consequently not capturing the jobs, tax revenues and other economic benefits which derive from technological development and construction. As a result of these characteristics, enclaves operate almost as separate economic spaces from their host economies and are more linked to more developed ones overseas. This is sometimes graphically illustrated. Some of the oil rigs off the Equato-Guinean coast have Texas phone dialling codes (Maass 2009), and sometimes locals in West Africa refer to oil rigs as mosquitoes as they extract oil (blood) and often bring problems (disease) in their wake. All of this being the case, as currently constituted enclaves do not, for the most part, promote economic diversification and development and the limited linkages to local economies could also be considered part of the resource curse. This is particularly significant in sub-Saharan Africa where although there are relatively few substantial oil-producing countries, oil accounts for half of total exports, with much of the rest accounted for by other unprocessed primary commodities. Of course, the fact that many natural resource operations in the Global South are also foreign-owned means that profits flow overseas rather than being domestically reinvested, for the most part.

Socio-political channels

When the resource curse is operative, there are also social and political channels through which it flows. In resource-abundant economies political elites do not necessarily have an incentive to try to diversify and upgrade production structures. Rather, taxes and rents can be raised from often enclaved resource sectors, often making political elites more interested in relations with TNCs than with the local citizenry.

Transparent, effective and broadly responsive governments have social contracts with their populations: that is, populations agree to pay taxes in return for social services, such as health and education, and security. However, in resource-based economies there is often a different type of social contract between domestic political elites, TNCs and world powers: what I have called elsewhere a transnational contract of extraversion (Carmody 2010). An extraverted economy is one which is oriented to serve the needs of people overseas, rather than the majority of its own residents through the provision of raw materials or fuel and as a market for manufactured goods and services from elsewhere. This was largely what colonialism was about – trying to turn the terms of trade in favour of the colonising countries and against the colonies.

The "resource curse" then should be analysed in historical perspective. Resource dependence for many countries is partly an outcome of their deliberate deindustrialisation under colonialism. For example, when the British East India Company colonised India, Britain taxed and then banned imports of certain types of textiles (calico) from that territory to favour its own industry. There were also contemporaneous reports that the British at times cut off weavers' thumbs, although the evidence for that is disputed. In the 1760s the British prime minister, Pitt the Elder, said that the colonies should "not be permitted to manufacture as much as a horseshoe nail" (quoted in List 1916, 55 in Raudino 2016, 111).

Socio-economic conditions in Indian cities deteriorated to such an extent under British rule that the colony at one point ruralised – that is, many people fled the cities in an attempt to at least subsist in the rural areas (Stavrianos 1981).

> In 1700, before colonialism, the United Kingdom had 2.9 per cent of world GDP and India 24.4 per cent. Shortly after independence India's proportion of world GDP had dropped to just 4.2 per cent while the UK's had risen to 6.5 per cent, down from a high of over 9 per cent in the late nineteenth century.
>
> (Brooks 2017, 72)

When the British led the "liberation" of Ethiopia from Italian occupation during World War II they largely dismantled its manufacturing because it was seen to be inappropriate for the stage of development of its people. According to the Eritrean Ministry of Information (2009) website "after the war in Libya was over [after the Germans had been defeated], Britain shifted its base to other colonies and transferred out most of the modern manufacturing companies in Eritrea".

Under colonialism there were even prohibitions on the growing of certain types of crops. For example, in 1751 the British Board of Trade gave an order to the Cape Castle, the Centre of British Administration in the Gold Coast [now in Ghana], to stop cocoa cultivation amongst the Fante, saying that:

> The introduction of culture and industry among the Negroes is contrary to the known established policy of this country, there is no saying where it might stop, and that it might extend to tobacco, sugar and every other commodity which we now take from the colonies and thereby the Africans, who now support themselves by wars, would become planters and their slaves be employed in the culture of these articles in Africa, which they are employed in America.
>
> (quoted in Boahen 1966, 113 in Wallerstein 1986, 14)

After the end of the slave trade the British assisted their West African colonies to produce cash crops, although production of manufacturing was still prohibited, as these items continued to be imported from the "mother country".

The structure of the transnational contract of extraversion is associated with extreme income inequality. For example, perhaps the most iconic example of the resource curse is Nigeria, where the majority of the population has gotten poorer since independence, while the World Bank estimates that 80 per cent of oil revenues that remain in Nigeria are captured by just 1 per cent of the population (Ekong *et al.* 2013). The oil-bearing communities in the Niger Delta suffer the burden of massive pollution, with the equivalent amount of oil from the Exxon Valdez tanker disaster in Alaska being spilt every year into the region (Klein 2014), largely arising from pipeline oil theft and sabotage. This practice of breaking into oil pipelines to steal their contents is known as "bunkering".

Meanwhile local elites and transnational company shareholders profit from their resources. This has given rise to a conflict complex in the Niger Delta. A conflict complex arises when war becomes a form of social equilibrium. The conflict in the Niger Delta seemed to be brought under control when the Nigerian government paid militants stipends to remain demobilised; however, as these were cut back by the new government in 2016, the conflict resumed, with a new militant group – the Niger Delta Avengers – attacking oil pipelines and installations. In some cases peaceful protests have been violently repressed by the Nigerian state. The execution of the poet Ken Saro-Wiwa and others from the Movement for the Survival of the Ogoni People in 1995 who advocated non-violence by the Nigerian state was implicated in the rise of violent militia/resistance groups in the region.

Sometimes conflicts such as the one in the Niger Delta are called "resource conflicts", although in reality this is a misnomer, as human conflict is always socio-political, with resources and the distribution of rents from them sometimes providing a source of grievance and of funding to buy arms, for example. Such types of conflict can be aggravated by what is sometimes known as horizontal inequality. Vertical inequality is between individuals or households, whereas horizontal inequality refers to income differences between groups. Where the benefits of resource rents are monopolised by particular ethnic groups, for example, this may further fuel resentment and conflict.

Socio-political, economic and environmental factors can interact to deepen the resource curse. As people get poorer or are exposed to environmental degradation, they may feel aggrieved, and if their livelihoods are undercut, then what economists refer to as the opportunity cost of conflict may be reduced. If someone has a good job or livelihood, then the opportunity cost of them engaging in armed conflict is high, as they would lose their job. On the other hand, if they are unemployed, as many youth in the Niger Delta are, then the opportunity cost of being in a militia is low, as this may provide a source of income for them which they cannot garner from other sources, given pervasive poverty and unemployment.

Given high levels of vertical and often horizontal inequality associated with the transnational contract of extraversion, it is not surprising that it is associated with high levels of conflict. In reality, the resource curse is a mode of governance which generates immense wealth for some and poverty and exclusion for many. It can be conceived of as an outcome of a globalised hydrocarbon assemblage or network of people and things, which manifests differentially depending on pre-existing local and national conditions (Siakwah 2017b).

Oftentimes in the popular Western imagination there is implicit methodological nationalism. It is not uncommon to hear people say, for example, that Nigeria is poor because of corruption – thereby discursively isolating the problem to the territory of the country and its political elites. However, corruption often requires both a payee and a payer, who are often transnational oil companies. For example, Halliburton, the American oil service company, agreed to pay the Nigerian government tens of millions of dollars to settle corruption cases against it, including charges against former vice president of the US, Dick Cheney, who had previously been its chief executive officer (CNN Wire Staff 2010).

Comaroff and Comaroff (2012, 15) argue that

> because market forces in Africa have never been fully cushioned by the existence of a liberal democratic state and its forms of regulation, and . . . because governance there has frequently been based on kleptocratic patronage – all of these things also being in part legacies of colonialism and its aftermath – African polities have been especially hospitable to rapacious enterprise: to asset stripping, to the alienation of the commons to privateers, to the plunder of personal property, to foreign-bribe giving. In sum, to optimal profit at minimal cost, with little infrastructural investment.

(Geo)political channels

> Afghanistan may be the graveyard of empires, but Iraq is home to a graveyard sense of humor. Iraqis wonder aloud whether the U.S. and Britain would have invaded Iraq if its main export had been cabbages instead of oil.
>
> – (McGovern 2011)

Another of the channels through which resource abundance may negatively affect a country is through (largely) unwanted attention or intervention from outside powers. The perhaps iconic example of this is the 2003 American-British invasion of Iraq. Prior to this Iraq was ruled by a despotic dictator, Saddam Hussein. However, when he was deposed what followed was a brutal and long-lasting civil war, which was largely responsible for spawning the Islamic State terrorist movement, which, in turn, caused huge suffering, devastation and destruction in Syria and elsewhere. It has now been largely accepted that oil played a significant role in the American decision to invade Iraq, as there were no weapons of mass destruction in that country prior to the invasion (Ritter and Pitt 2002). Scott Ritter was a United Nations weapons inspector in Iraq prior to the invasion, who was later the subject of a "sting" operation in the US. The idea of invading parts of the Middle East to control its oil supply had been circulating in American policy circles for decades (North 2016).

The fact that the attack on Iraq was largely conducted for geopolitical and geo-economic rather than security reasons can be seen in the chronology of events. The day after the terrorist attacks of 11 September 2001 in the United States, Secretary for Defence "Rumsfeld complained there were no decent targets for bombing in Afghanistan [where the attacks were orchestrated from] and we should consider bombing Iraq, which had better targets" (Clarke 2004, 31). Bombing a country because it has good targets, such as power stations, would appear to be a strange logic. By that reasoning the United States could bomb itself. Indeed, in response to Republicans questioning the legality of targeted assassination of American citizens by drones overseas, the US attorney general under Obama did issue an opinion that it would be legal for the US government to also do this on American soil.

There have also historically been other examples where the United States and/or other Western powers have been involved in deposing heads of other resource-rich states, such as Mossadegh in Iran and Patrice Lumumba in Zaire, and mostly recently Gadaffi in Libya in 2011, again plunging that country into chaos, with a slave trade in sub-Saharan African immigrants trying to get to Europe via Libya exposed in 2017. In Zaire (now the DRC) the assassination of Patrice Lumumba in 1961 paved the way for the ascent to power of Joseph Mobutu, who oversaw both massive personal wealth accumulation for himself and economic and social retrogression and degradation for the population at large.

Gaining access to raw materials and markets is a central objective for developed countries in their interactions with the Global South. In the case of the European Union, this can be seen through its "Economic Partnership Agreements" with countries in Africa, which seek to open up export markets (for more details see Carmody 2016), and through its Raw Materials Initiative (European Commission 2008). Another mechanism is through SAPs of the World Bank and IMF, which through trade and investment liberalisation and privatisation open up countries' markets and resources to access by outside companies, which are often TNCs. We can see the coming together of these different types of power projection through the idea of matrix governance. The concept of matrix governance shares similarities with Gramsci's idea of hegemony, defined as coercion informed by consent (Marais 1998). However, given its globalised operation, matrix governance places less emphasis on culture as conventionally understood and more on assemblages of power and their operation across space.

Matrix governance, ideology and the resource curse

As Sassen (2014, 84) notes in relation to the operation of the IFIs and the WTO, "the resulting mix of constraints and demands had the effect of disciplining governments not yet fully integrated into the regime of free trade and open borders pursued mostly by large firms and the governments of dominant countries". Matrix governance attempts to prescribe and proscribe certain sets of actions for developing country states and populations to enable "unimpeded" accumulation of and by TNCs. Indeed, the process of globalisation could be thought of as the global deregulation of capital through the international regulation of states (Carmody 2002). Thus, the objective of matrix governance is to promote order, sometimes through the threat or application of force, and a matrix for globalised capital accumulation which creates strong property rights and insurance against expropriation, for example. The structure of interlocking treaties, institutions and agreements circumscribes the actions available to states while widening the latitude of action of TNCs. However, this is not without its resistances in the form of different types of social movements, from 'anti-globalisation' to terrorist ones.

Some Islamic terrorist movements have a different vision for the type of global state they want – a global Islamic state, where sovereignty belongs to God, not people. The reasons behind these movements are complex and vary by context; however, they are united by a disaffection with the current exercise of sovereignty by the states which are notionally in control, if not always fully effectively, of territories in which they operate. These movements seek to establish new forms of governance based on sharia law, rather than the imperatives of profitability and accumulation.

According to Michael Watts (2004) resource extraction often generates ungovernability, although there are also counter-examples such as Norway. This may operate through a variety of channels. As noted earlier, the ecological and social catastrophe in the Niger Delta, where Nigeria's oil is produced, has given rise to a variety of resistance movements. However, the marginalisation of the north and northeast of the country, partly as result of the uneven flow and distribution of oil revenues and macroeconomic effects, has also recently been implicated in the rise of the terrorist group Boko Haram. This movement seeks to establish an Islamic caliphate in the northeast of the country and more broadly (Smith 2015). It seeks a new modality of governance, distinct from what it sees as the corrupt and degenerate current nation-state. According to Lee (2014) neoliberalism hybridises with other forms of governance, such as neopatrimonialism, to create Afro-neoliberal capitalism. This Afro-neoliberal capitalism is arguably the target of Boko Haram, although some of its leaders have also been known to own luxury marque cars. In this case then (fundamentalist) ideology may also be used to serve political ends. The creation of uneven development, which matrix governance exacerbates and attempts to stabilise, is implicated in the generation of vertical and horizontal inequalities, which interact with identity in particular ways, which may, in turn, generate violent resistance.

Islam has for centuries served as an axis of resistance to Western domination, from the Crusades to the Mahdi in Sudan in the nineteenth century, to today. It can serve as a basis of identity for people who have been declassed, discriminated against or subjected to the dislocations of globalisation – expelled in Saskia Sassen's terminology.

Islamic fundamentalist ideology seeks a different moral ordering where the lives of the ummah (Islamic community) are structured, ordered and constrained by a set of principles which are considered to be divinely ordained and consequently just – 'answering only to God' (Abdo and Lyons 2003). However, those who are considered to be non-believers are either to be converted under the sign of the sword or eliminated. The authoritarianism of matrix governance thus achieves its analogue in the authoritarianism of Islamic fundamentalism, as they partly co-constitute, and the role of resources is often central in these social relations.

The marginalisation of youth, which is often generated by globalisation (Sommers 2015), results in a destabilisation and search for new sources of identificatory attachment. In a sense Islamic rebellions around the world could be seen as struggles over the territorialisation of the global capitalist state, as Western powers attempt to use "lawfare" (Morrissey 2011), amongst other techniques, to regulate and stabilise space. Lawfare is a mix of warfare and law. This would include things such as enhanced surveillance or sometimes bending, breaking or abrogating of laws, such as designating people as "unlawful enemy combatants" so that they can be held off-shore in the US military base in Guantanamo Bay in Cuba, so that they can be denied access to recourse through the American legal system. There is a global civil war now being fought over different visions of the type of global state which should be constructed and embedded.

Materiality, lootability and "resource conflicts"

Karl Marx famously argued against the idea of commodity fetishism – that is, the idea that through the medium of money in a market economy, there appear to be relations between things, so corn can be sold to buy oil, for example. According to some analysts oil and credit constitute the twin pillars of modern "petro-market civilization" (DiMuzio 2011). Money and the equivalences it establishes between different things serve to disguise the social relations of production and exchange, and the same can be said about fetishising oil – giving it independent power, whereas in reality:

> Oil comes to mark a particular epoch (like the age of coal or steam) and to this extent is not only a bearer of particular relations of production but it is equally a source of enormous political and economic power and therefore it carries a set of ideological and cultural valences as is implied in the moniker of 'black gold' or 'petro-dollars' (it is both a commodity and a commodity fetish). In this account oil (and other key resources) has causal powers: it is a purveyor of corruption, it undermines democracy, promotes civil and inter-state wars ('blood for oil'), is the mother forms of corporate power ('Big Oil') and condemns oil-rich states to devastating economic, political and social pathologies (oil is the 'devil's excrement' as a former head of OPEC once put it).
>
> (Watts 2009, 14)

One of the academics who has made the biggest contributions to understanding resource-related conflicts is Philippe Le Billon, who is a professor of geography at the

University of British Columbia in Canada and has written seminal articles and books on the subject (Le Billon 2004; Le Billon 2008; Le Billon 2012). In these writings he cautions us against attributing independent causative powers to resources and insists on analysing the ways in which resources are inserted into pre-existing socio-political configurations that will influence or determine outcomes. Thus, to designate something as a "resource conflict" may occlude as much as it illuminates.

Le Billon gets us to think about how the geography of resources matters. He distinguishes between point resources, such as oil, and geographically diffuse resources, such as fish or forests. He also argues that the extent to which resources can be used by rebel groups to fund their activity also depends on their lootability. For example, Kimberlite diamonds, named for the town in South Africa where there was a diamond rush in the nineteenth century, are found in the rock of the same name found deep underground. This means that in order to extract them, there has to be capital-intensive investment to dig the mine. This also requires infrastructure and a stable political environment. Consequently, Kimberlite diamonds are not lootable, at least not before they come out of the ground. This is not the case for so-called arable or alluvial diamonds, which are found in soil or in the beds of rivers. These were originally Kimberlite diamonds, as high pressure is required to turn carbon into diamond, but then they have been excavated through natural erosive processes. All that is required to extract this type of diamond is basic equipment such as a pick and shovel or pan. This makes them more easily lootable, and consequently they have frequently been used by rebel groups such as the Revolutionary United Front (RUF) in Sierra Leone to fund their activities. The RUF's signature punishment was to cut people's hands off. They reportedly did this because the hand was a symbol of the governing party, to which they were opposed, and it prevented people from voting. It is also reported that it was prompted by the desire to reduce the number of miners available to the government to mine diamonds. While such practices are often presented as an example of supposed "African barbarism" and the "coming anarchy" associated with "tribalism" (Kaplan 1994), it is important to remember that it was the Belgians who introduced the practice of cutting off peoples' hands in the Congo when they were judged not to have tapped enough rubber.

Point resources are frequently more lootable than diffuse ones and are found in smaller areas that are easier for rebel groups to control, although there are also counter-examples. While many people are familiar with the idea of "blood diamonds", often as result of the eponymous film starring Leonardo DiCaprio, sometimes rebel groups or governments force people to cultivate cocoa, for example, resulting in "blood cocoa", which sometimes works its way into our global chocolate supply. Just three TNCs – Cargill (USA), Barry Callebaut (Swiss) and Archer Daniels Midland (USA) – between them process over 40 per cent of the global cocoa supply (Ryan 2011).

Transcending the resource curse?

The resource curse is a highly contentious topic, with some claiming that it does not exist and that in actuality it is a result of the creation of distorted economies created by colonialism (Wengraf 2017). Others argue that because some natural resource-rich countries have not experienced a curse, such as Norway (oil) or Botswana (diamonds) in Southern Africa, that it is in reality a governance, not a resource curse. Norway, for example, has invested hundreds of billions of dollars of oil revenue in the world's largest sovereign wealth fund, much of which is invested overseas so as not to "over-inflate" the domestic economy. This is one of the ways in which massive inflows of foreign currency into the economy can be "sterilised". Interestingly in 2013 this Norwegian fund divested from 23 palm oil companies, which are responsible for substantial deforestation, and 27 mining companies (Reyes 2016).

There is, however, evidence that resource abundance in the absence of a diversified economy with good institutions, such as in the UK or Norway, does tend to create problems for development, depending on the context. Michael Ross (2012, 3), in his magisterial work on the "oil curse", found that

> analysing 50 years of data for 170 countries in all regions of the world, it finds little evidence for some of the claims made by earlier studies: that extracting oil leads to abnormally slow economic growth, or makes governments weaker, more corrupt, or less effective. On some fronts, like reducing child mortality, the typical oil state has outpaced the typical non-oil one.

However, since 1980 he finds that oil-producing countries have become less democratic and are more likely to suffer from violent insurgencies. Furthermore, he argues that while they have grown economically by about the same rate as non–oil-producing countries during that time. They should have grown faster given their natural resource abundance.

Partly the impact of resources depends on the scale of the deposits, the nature of governing institutions and population size. Qatar, the oil-rich state in the so-called "Middle East", has one the highest per capita incomes in the world according to the World Bank (2017c). The famous post-colonial theorist Edward Said objected to the term "Middle East" given its Eurocentric construction, asking what it was in the middle of and east of where? The answer to the east of where is the Prime Meridian in Greenwich in England, and Britain's geopolitical power allowed it to establish the division of the world geographically through the development of the system of lines of latitude and longitude.

Qatar is ranked as the eleventh-richest country in the world in purchasing power parity terms, although many of the states ahead of it are so-called "micro-states" with small populations which serve as tax havens such as the Channel Islands and Monaco. According to some analysts it is the richest country in the world, with an average per capita income of US$129,726 (Gregson 2017). The lengths to which many high-net-worth individuals will go to avoid paying tax are illustrated by the fact that it is reported that some of them hire people in Monaco to switch lights on and off and run the water in their apartments to make it look like they are residing there for tax purposes (Monaco has no personal income tax). There is also a cruise ship called "the World", where such people can buy apartments for millions of dollars. It sails around the world incessantly, meaning its residents are not located anywhere for tax purposes (Urry 2014).

Qatar has the fourteenth-largest proven oil and gas reserves in the world according to the United States Energy Information Administration, but a population of only around two and a half million people (Energy Information Administration 2017). The scale of oil revenue, given the small population, has been transformative for the economy (Newman *et al.* 2016) and allowed the state to set up its own airline, Qatar Airways, which flies to over 70 countries worldwide giving it the fifth-largest international route network in the world (Garfors Globe 2013). The huge inflows of foreign currency generated by oil in Qatar mean the country, or its citizens, have investments of more than US$30 billion in London alone in the department store of Harrods, Claridges Hotel and the London Stock Exchange and in the iconic office developments in Canary Wharf and "the Shard", for example. Reportedly the Qatari royal family owns more of London than the British Crown Estate, which manages Crown holdings not held privately. In 2017–2018 there was a diplomatic crisis in Qatar, as a Saudi-led coalition imposed sanctions on the country, accusing it of supporting terrorism. As of 2018 the Gulf Cooperation Council was still economically blockading Qatar as a result of these accusations.

A similar argument could be made for Botswana in Southern Africa not succumbing to the resource curse, which despite having a large land area, has a population of only 2 million people or so. There the fact that it was regarded as a backwater in colonial times may have been an advantage, as extractive institutions and practices were not put in place in the same way as they were elsewhere in much of Southern Africa. Furthermore the government of Botswana has been dominated historically by large cattle ranchers, who had a common interest in promoting institutions and conditions favourable to business and capital accumulation (Samatar 1999).

A variety of solutions have been put forward to overcome the resource curse. These include policy proposals and the creation of institutions, such as the Publish What You Pay initiative and the Extractive Industries Transparency Initiative (EITI), which was headed by the former secretary for international development in the UK, Claire Short, until 2016. Both of these initiatives aim to promote transparency in natural resource governance by getting companies and governments to release details of their contracts. There is also the Natural Resources Charter, which propounds principles for good natural resource governance, which the Oxford economist Paul Collier was involved in designing. While improvements may be possible through these types of initiatives if governments sign up for them, a general problem is that telling political elites what they should do is often not productive, as they often already manage natural resources in ways they perceive to be to their own benefit.

The resource curse is a tendency, not a law (Auty 1993). This tendency can be counteracted or obviated through both macro-economic management and micro-economic policy, such as local content policies in the mining industry (Morris *et al.* 2012), which encourage the development of manufacturing to supply mining equipment. However, as noted earlier, the resource curse is actually a mode of governance which benefits powerful stakeholders and, as such, they have little incentive to change it. Ultimately, the resource curse will only be overcome through national and transnational struggles for state accountability.

The resource curse arguably has its origins in the deindustrialisation wrought by colonialism (Wengraf 2018), much of which took place in cities. What role do cities play in development? As the majority of the world's population is now urban, this is a pressing question, which will be explored in the next chapter.

Further reading

Books

Auty, R. 1993. *Sustaining Development in Mineral Economies: The Resource Curse Thesis*. London: Routledge.

Humphreys, M., J. Sachs, and J. Stiglitz, (eds.) 2007. *Escaping the Resource Curse*. New York: Columbia University Press.

Wengraf, L. 2018. *Extracting Profit: Imperialism, Neoliberalism and the Scramble for Africa*. Boston: Haymarket Books.

Articles

Sachs, J., and A. Warner. 2001. "The Curse of Natural Resources." *European Economic Review* 45: 827–38.

Watts, M. 2004. "Antimonies of Community: Some Thoughts on Geography, Resources and Empire." *Transactions of the Institute of British Geographers* 29: 195–216.

Websites

Extractive Industries Transparency Initiative, https://eiti.org/

Natural Resource Charter, https://resourcegovernance.org/approach/natural-resource-charter

References

Abdo, G., and J. Lyons. 2003. *Answering Only to God: Faith and Freedom in Twenty-First-Century Iran*. New York: John Macrae Book.

Auty, R.M. 1993. *Sustaining Development in Mineral Economies: The Resource Curse Thesis*. London: Routledge.

Boahen, A. 1966. *Topics in West African History*. London: Longmans.

Brooks, A. 2017. *The End of Development: A Global History of Poverty and Prosperity*. London: Zed Books.

Campbell, C. 2003. *Letting Them Die: Why HIV/AIDS Intervention Programmes Fail*. Oxford: International African Institute in Association with James Currey.

Carmody, P. 2002. "Between Globalisation and (Post)Apartheid: The Political Economy of Restructuring in South Africa." *Journal of Southern African Studies* 28 (2): 255–75.

———. 2010. *Globalization in Africa: Recolonization or Renaissance?* Boulder, CO: Lynne Rienner Publishers.

———. 2016. *The New Scramble for Africa*. 2nd edition. Cambridge: Polity.

Clarke, R.A. 2004. *Against All Enemies: Inside America's War on Terror*. New York and London: Free Press.

CNN Wire Staff. 2010. "Halliburton Settles Nigeria Bribery Claims for $35 million." http://edition.cnn.com/2010/WORLD/africa/12/21/nigeria.halliburton/index.html.

Collier, P. 2011. *The Plundered Planet: How to Reconcile Prosperity with Nature*. London: Penguin Books.

Comaroff, J., and J. Comaroff. 2012. *Theory from the South, or, How Euro-America Is Evolving Toward Africa*. Boulder, CO: Paradigm Publishers.

DiMuzio, T. 2011. "The Crisis of Petro-Market Civilization: The Past as Prologue?" In *Global Crises and the Crisis of Global Leadership*, ed. S. Gill. Cambridge: University Press, p. 73.

du Venage, G. 2017. "South African Mines Dig Deep into Technology." *The National*. www.thenational.ae/business/south-african-mines-dig-deep-into-technology-1.37286.

Ekong, C., E. Essien, and K. Onye. 2013. *The Economics of Youth Restiveness in the Niger Delta*. Houston: Strategic Book Publishing & Rights Agency, LLC.

Energy Information Administration. 2017. "International Energy Statistics." www.eia.gov/beta/international/data/browser/#/?pa=000000000000000000000008&c=ruvvvvvfvtvnvv1urvvvvfvvvvvvfvvvou20evvvvvvvvvnvvuvo&ct=0&tl_id=5-A&vs=INTL.57-6-AFG-BB.A&cy=2016&vo=0&v=H.

Eritrean Ministry of Information. 2009. *British Adminstration (1941–45)*. www.shabait.com/about-eritrea/history-a-culture/591-britsh-adminstration-1941-45.

European Commission. 2008. "Communication from the Commission to the European Parliament and the Council – The Raw Materials Initiative: Meeting our Critical Needs for Growth and Jobs in Europe {SEC (2008) 2741}." http://eur-lex.europa.eu/legal-content/EN/TXT/?uri=CELEX:52008DC0699.

Garfors Globe. 2013. "These Airlines Fly to Most Countries." http://garfors.com/2013/03/the-worlds-most-international-airlines.html/.

Gregson, J. 2017. "The World's Richest and Poorest Countries." *Global Finance*. www.gfmag.com/global-data/economic-data/worlds-richest-and-poorest-countries.

Kaplan, R. 1994. "The Coming Anarchy: How Scarcity, Crime, Overpopulation, Tribalism, and Disease Are Rapidly Destroying the Social Fabric of our Planet." *The Atlantic*. www.theatlantic.com/magazine/archive/1994/02/the-coming-anarchy/304670/.

Karl, T. 1997. *The Paradox of Plenty: Oil Booms and Petro-States*. Berkeley, LA and London: University of California Press.

Klein, N. 2014. *This Changes Everything: Capitalism Vs. The Climate*. London: Allen Lane.

Le Billon, P. 2004. "The Geopolitical Economy of 'Resource Wars'." *Geopolitics* 9 (1): 1–28.

———. 2008. "Diamond Wars? Conflict Diamonds and Geographies of Resource Wars." *Annals of the Association of American Geographers* 98 (2): 345–72.

———. 2012. *Wars of Plunder: Conflicts, Profits and the Politics of Resources*. London: Hurst.

Lee, M. 2014. *Africa's World Trade: Informal Economies and Globalization from Below*. London: Zed Books.

Maass, P. 2009. *Crude World: The Violent Twilight of Oil*. London: Allen Lane.

Marais, H. 1998. *South Africa: Limits to Change: The Political Economy of Transition*. London: Zed Books.

McGovern, R. 2011. *Surprise, Surprise! Iraq War Was About Oil*. www.resilience.org/stories/2011-04-23/surprise-surprise-iraq-war-was-about-oil/.

Morris, M., R. Kaplinksy, and D. Kaplan. 2012. *One Thing Leads to Another: Promoting Industrialisatoin by Making the Most of the Commodity Boom in Sub-Saharan Africa*. Self-published.

Morrissey, J. 2011. "Liberal Lawfare and Biopolitics: US Juridical Warfare in the War on Terror." *Geopolitics* 16 (2): 280–305.

Næss, A. 1989. *Ecology, Community and Lifestyle: Outline of an Ecosophy*. Cambridge: Cambridge University Press.

Newman, C., J. Page, J. Rand, A. Shimeles, M. Soderbom, and F. Tarp. 2016. *Made in Africa: Learning to Compete in Industry*. Washington, DC: Brookings.

Nino, H.P., and P. Le Billon. 2014. "Foreign Aid, Resource Rents, and State Fragility in Mozambique and Angola." *Annals of the American Academy of Political and Social Science* 656 (1): 79–96.

North, D. 2016. *A Quarter Century of War: The US Drive for Global Hegemony*. Oak Park: Mehring Books.

Raudino, S. 2016. *Development Aid and Sustainable Economic Growth in Africa: The Limits of Western and Chinese Engagements*. Switzerland: Palgrave Macmillan.

Reyes, O. 2016. "Climate Change Inc: How TNCs Are Managing Risk and Preparing to Profit in a World of Runaway Climate Change." In *The Secure and the Dispossessed: How the Military and Corporations Are Shaping a Climate-Changed World*, eds. N. Buxton and B. Hayes. London: Pluto.

Ritter, S., and W. Pitt. 2002. *War on Iraq: What Team Bush Doesn't Want You to Know*. London: Profile.

Ross, M.L. 2012. *Oil Curse: How Petroleum Wealth Shapes the Development of Nations*. Princeton, NJ: Princeton University Press.

Ryan, O. 2011. *Chocolate Nations: Living and Dying for Cocoa in West Africa*. London: Zed Books.

Samatar, A.I. 1999. *An African Miracle: State and Class Leadership and Colonial Legacy in Botswana Development*. Portsmouth, NH: Heinemann.

Sassen, S. 2014. *Expulsions: Brutality and Complexity in the Global Economy*. Cambridge, MA: Belknap and Harvard University Press.

Siakwah, P. 2017. "Political Economy of the Resource Curse in Africa Revisited: The Curse as a Product and a Function of Globalised Hydrocarbon Assemblage." *Development and Society* 46 (1): 83–112.

Smith, M. 2015. *Boko Haram: Inside Nigeria's Unholy War*. London: I.B.Tauris.

Sommers, M. 2015. *The Outcast Majority: War, Development and Youth in Africa*. Athens, Georgia and London: Georgia University Press.

Stavrianos, L. 1981. *Global Rift: The Third World Comes of Age*. New York: Morrow.

Urry, J. 2014. *Offshoring*. Cambridge: Polity.

Wallerstein, I. 1986. *Africa and the Modern World*. Trenton, NJ: Africa World Press.

Watts, M. 2004. "Antimonies of Community: Some Thoughts on Geography, Resources and Empire." *Transactions of the Institute of British Geographers* 29: 195–216.

———. 2009. "Crude Politics: Life and Death on the Nigerian Oil Fields." *Niger Delta Economies of Violence Working Paper No. 18*. http://citeseerx.ist.psu.edu/viewdoc/download?doi=10.1.1.518.4318&rep=rep1&type=pdf.

Wengraf, L. 2017. "The Pillage Continues: Debunking the Resource Curse." *Review of African Political Economy*. blog. http://roape.net/2017/01/24/pillage-continues-debunking-resource-curse/.

———. 2018. *Extracting Profit: Imperialism, Neoliberalism and the Scramble for Africa*. Boston: Haymarket Books.

World Bank. 2017. "Gross National Income Per Capita 2016, Atlas Method and PPP." http://databank.worldbank.org/data/download/GNIPC.pdf.

CHAPTER

10

Urbanisation and development

Generative cities or slumification?

Globalisation is often thought to be a relatively recent phenomenon of the last few decades. In reality, there is a much longer history of inter-connection between people and places in different world regions. European colonialism was a form of globalisation which initially took the form of invasion of other territories and war (Barkawi 2006), which was then followed by imposed patterns of economic restructuring. The British colonial aphorism that "trade follows the flag" captures this relationship well, and cities played a key role in the development and administration of colonies in the Global South (King 1990).

For the first time in our history humanity is now primarily an urban species. Fifty-four per cent of the globe's population now live in cities, and this is projected to rise to 66 per cent by 2050 (United Nations, Department of Economic and Social Affairs, Population Division 2014). The number of people living in cities around the world grew very rapidly from around three-quarters of a billion in 1950 to almost 4 billion in 2014. What this means is that what happens in cities will affect most of the world's population and that "development" is becoming an increasingly urban question, even if most of the worst poverty in the world is still in rural areas.

What is a city? The urban and development

Urban studies is a distinctive and well-developed field of its own; however, for our purposes a number of vitally important questions have to addressed if we are to think about and get purchase on the role of cities in development. First, what is a city? This may seem like a silly question – surely we know cities when we see them? They are typically characterised by dense population settlement, built up and populated by people who are primarily engaged in non-agricultural activities. However, what exactly constitutes a city or the urban space is still subject to dispute and interpretation. For example, with the severe economic crisis across much of Africa in the 1980s and 1990s many people took to urban agriculture in order to supplement their incomes and food sources as part of the burgeoning informal sector (Hampwaye *et al.* 2007). Although this is also happening in some Western cities, such as Detroit in the US, where in some cases entire city blocks have been demolished and the space devoted to this activity, thereby showing the globalised nature of economic restructuring.

Cities are partly created through their interactions with other places, through flows of people or raw materials, for example, and as such are "translocalities" (Appadurai 1995). Whereas in Western contexts people often tend to think of the rural and urban as separate, in other parts of the world many people often maintain strong family linkages with rural areas of origin and may not consider themselves to be primarily urban dwellers, but maintain patterns of circular migration between the urban and rural (Potts 2010). In the case of Zimbabwe, its recent economic crisis meant that many people blurred the boundary between the urban and the rural as they sought to put together multiple modes of livelihood (Mustapha 1992) which spanned both (Bryceson and Mbara 2003).

Recently, within geography, there has been a debate about the ontological status of cities. Neil Brenner (2015, 15) has questioned "established understandings of the urban as a bounded, nodal and relatively self-enclosed sociospatial condition in favour of more territorially differentiated, morphologically variable, multiscalar and processual conceptualizations". Malaquais (2007, 32) argues that "it would be significantly more productive to discuss cities more generally, with given African cities as starting points, as prototypes for an emerging, global form of urbanity", whereas Simone (2004) writes of the "transterritorial city".

Although urbanisation is sometimes presented as a development challenge for the Global South, the process itself can bring benefits. Some, such as the World Bank (2009), see cities driving the development process around the world. According to the Bank (8) "growth comes earlier to some places than to others. Geographic differences in living standards diverge before converging". The majority of world economic output is disproportionately produced in cities. For example, Sao Paulo in Brazil has less than a tenth of the country's population but produces more than a third of its economic output (Friere and Polése 2003 cited in Beall and Fox 2009). There are a variety of different types of what economic geographers call "agglomeration" and urbanisation economies which arise from businesses and people being located close to each other in cities, such as access to suppliers and labour.

There are differences between types of urbanisation taking place through time and space. Whereas urbanisation was associated with industrialisation in the West and some parts of the developing world, this is not the case in much of Africa. This creates a variety of challenges. In the first instance, there is the issue of employment. Most new migrants find work in the so-called "informal sector". Indeed, given severe rural crises in many parts of the world, the poor have been more rapidly urbanising than entire populations (Ravillion *et al.* 2008 cited in Beall and Fox 2009). Thus, informal urbanisation is one of the most important channels of the broader process.

Again, there are a variety of definitions of what constitutes the informal sector, but it is generally considered to be the unregulated sector of the economy where people do not pay taxes or have licences to operate businesses, etc. It ranges from people being individually employed – selling tomatoes, for example – to large-scale operations running fleets of taxis. In some cases the informal sector can be an important source of revenue for state elites and officials, as they have the ability to "arbitrage" between it and the formal sector by taking bribes for not charging import duties, for example.

In some countries such as India up to 90 per cent of the labour force are employed in the informal sector (Selwyn 2017). Many workers in the informal sector also live in so-called "informal" housing, often in slums. The nature of work in the informal sector is problematic for a variety of reasons. In the first instance, the informal sector operates more like a "free" market based on survival of the most powerful, or the "law of the jungle" (Castells 1994). Because there are no formal contracts in the informal sector, it exposes people to exploitative labour practices by their employers, unless they are self-employed, which many people are. It could be argued that many people in the informal sector are also forced to engage in self-exploitation. Also, as noted earlier, the marginal productivity of labour in the informal sector is low. Adding another person selling tomatoes by the side of the street does not raise national economic productivity, but may depress returns to other tomato sellers. It does, most importantly though, allow people to survive.

Writing in the 1970s, Milton Santos (1979) argued that rather than being a temporary aberration, the informal sector was functionally articulated and reproduced through its interaction with the formal economy, for example, through the provision of cheap wage goods consumed by formal-sector workers.[1] Thus, rather than the informal economy being reflective of exclusion, the so-called dual economy was functional to the broader demands and structural nature of (underdeveloped) capitalist accumulation (Portes and Walton 1981; Castells and Portes 1989). Much of the population of cities in the Global South are then both included and excluded along different dimensions. They may be simultaneously excluded from public services and included in the consumption and production dynamics of the informal economy and global capitalism. This urban structure of inclusionary exclusion is itself then reflective of the broader politico-economic dynamics, often of resource export dependence, which in turn is a reflection of the way in which much of the Global South has historically been subordinately incorporated into the global economy.

Urbanisation leads to economies of scale in the provision of services like health, water, education, electricity and others. Productive industries also benefit from concentration of their suppliers and consumers, and cities reduce transportation and communication costs. Cities serve commercial, and often administrative, functions and are places where goods and services are produced and consumed, and are consequently generally centres of economic growth. They also generally have substantial labour markets and more capital than rural areas, and consequently may sometimes be centres of technological innovation, although growing populations, in general, are sources of new ideas (Boserup 1993). However, some have questioned whether the relationship between urbanisation and economic growth holds in Africa, arguing its cities are not properly integrated into the global economy, but rather serve an entrepôt (or entry point) function as centres for the exports of primary goods and imports of manufactures. This means many African countries are experiencing urbanisation without industrialisation. Inadequate infrastructure and services, and sometimes lacklustre city management, may reduce economic growth in these cities, although there have also been noted cases of urban revival in Lagos, Nigeria or Lubumbashi in the DRC in recent years. Lagos's economy is bigger than that of East Africa's biggest one, Kenya, and would be Africa's fifth largest if it was compared with countries. Cape Town has also been noted by some as having effective city management, although the administration there has recently been subject to political infighting. Sometimes local developmental city states emerge when the opposition is in control, as they perceive an existential threat from the national government to their rule, and consequently focus more on effective service delivery and tax collection (Cheeseman 2018). In general, however, rapid urbanisation in Africa is associated with slum development, with over 40 per cent of city populations below the poverty line, with many of those over 50 per cent (UN-Habitat 2008 cited in Carmody and Owusu 2016). Urbanisation refers to the proportion of a country's population living in cities, whereas urban growth refers to the number of people living in cities. If the number of people living in cities is increasing rapidly, then there is rapid urban growth. So technically it is possible to have rapid urban growth without urbanisation if populations in rural areas are growing equally fast.

Slums or informal settlements account for nearly all urban growth in some rapidly expanding African cities. Slums in Nairobi, Kenya's capital, account for approximately a quarter of the city's population, which is thought to be around 4 million people (UN-Habitat 2013 cited in Carmody and Owusu 2016). Kibera, a slum neighbourhood in the city, is often cited as Africa's largest slum, although there is some dispute over this. One of the main drivers of slum growth are developments in rural areas. Arrighi (2000 cited in Selwyn 2014) estimates that approximately 65 per cent of the growth of the global urban population is as a result of rural-urban migration, partly as a result of land expulsions and grabbing.

Planetary urbanisation

As noted earlier, there has been a substantial debate in urban studies and geography in recent years about what constitutes a city. However, there have also been other vibrant debates about the nature, extent and impact of urbanisation. One of the theories which has received a lot of attention recently posits the existence of planetary urbanisation. This theory has its roots in Henri Lefebvre's idea that the whole world is becoming urban. According to Neil Brenner (2015, 15) many previous urban theories have suffered from "methodological cityism", that is, the assumption that cities are "settlement types characterized by certain indicative features (such as largeness, density and social diversity) that make them qualitatively distinct from a non-city social world (suburban, rural and/or 'natural') located "beyond" or "outside' them". Brenner, however, disputes this and sees cities as not having definite and clearly defined boundaries. Rather, as the world urbanises, what happens in cities increasingly determines or extends into what happens in rural areas around the world through the creation of demand for palm oil, for example, which results in new plantations being set up in countries such as Indonesia (Li 2007). Often, these new plantations are planned from "global cities", such as Singapore, where Wilmar Corporation, one of the world's largest palm oil producers, is located (Carmody and Taylor 2016). In this way urbanisation is what the well-known sociologist Anthony Giddens (1990) might call "distanciated". Brenner and Schmid

(2015) argue that urbanisation entails three mutually constitutive moments: concentrated, extended and differential.

As Kanai (2014a, 1082) argues "urbanization becomes a much more encompassing notion when understood in terms of functional articulation and extensive spatial impacts rather than in terms of commonly used criteria of minimum density thresholds, land cover and land-use typologies". This may be true, but encompassing is not the same as accurate. Another way to think of these relationships is that what happens in cities influences, but may not determine, what happens in rural areas (Storper and Scott 2016). This varies by context, and likewise what happens in rural areas may affect what happens in cities. For example, it is often said in Africa that power is gained in the countryside (where the majority of the population live) and lost in the cities (Boone 2014). In another example, industrialisation in Dongguan in China was led at the village rather than city level (McGee *et al.* 2007). So to speak of planetary urbanisation as a state, rather than an ongoing tendency or process, is perhaps mistaken.

Neoliberalism and globalisation of cities in the Global South

Globalisation and the spread of neoliberalism have drastically shaped the way cities are developing across the globe. Neoliberalism in much of Africa and Latin America was imposed through SAPs in the 1980s. These essentially transformed the state's role into a nightwatchperson of the "free" market, tasked with creating a stable investment climate and managing those marginalised by economic reforms (Peck and Tickell 2002; Afenah 2009 cited in Carmody and Owusu 2016). As noted earlier, globalisation refers to worldwide integration of economies and increased interaction amongst people, companies and governments which intensified from the late 1980s with the collapse of the Soviet bloc. This enhanced globalisation has, in part, been driven by international trade and investment and aided by ICT that make the world more integrated, and therefore more interdependent (Murphy and Carmody 2015).

The world economy is largely organised around and through so-called global or world cities, where the headquarters of TNCs and banks are based. As cities have globalised, and arguably driven the process, they aren't merely places where production and the exchange of goods and services takes place. Instead, they are increasingly places where products and people link into the wider world (Robinson 2006). Consequently, today's globalised urban system is dominated by a relatively small number of centres that serve as command-and-control points for global capitalism, such as New York, London and Shanghai.

Although most global or world cities are located in developed countries (Robinson 2006), many cities in the Global South are becoming central to capital accumulation globally, as centres of both production and finance, such as Shanghai, where one of China's main stock exchanges is based. Also a city's global importance may vary by sector. For example, Johannesburg in South Africa is an important centre of ownership globally in the platinum industry (Surborg 2011).

Rapid urbanisation and industrialisation in Asia have often resulted in environmental degradation, however.

> The World Health Organization sets the guidelines for the safe presence of fine particles of dangerous air pollutants (known as PM2.5) at 25 micrograms or less per cubic metre; 250 is considered hazardous by the US government. In January 2014, in Beijing levels of these carcinogens hit 671.
>
> (Klein 2014, 351)

While much of the literature has centred on global and world cities (GaWC), some have argued this focus perhaps occludes as much as it illuminates. As the planet urbanises, much of the world lives in "ordinary" cities (Robinson 2006). Indeed, more than half of the world's urban population live in cities of fewer than 500,000 people (UN-Habitat 2006). Myers (2018) notes that in the index produced by UN-Habitat 2006 the GaWC group includes only one African city (Johannesburg) recorded in the top "Alpha" level. Kinshasa, the capital of the DRC, with a population of approximately 12 million people, only entered the index in 2016 under the lowest "sufficiency" ranking. The occlusion of these types of cities from study, Robinson (2006) argued, set policy makers across the world on the wrong course, as they set out to make their urban centres "world cities", rather than addressing the urgent basic needs of the majority of the population.

Cities in the Global South also serve as new markets for Northern and other TNCs through so-called "bottom of the pyramid" strategies, which seek to consumerise the poor by selling small packets of washing powder, for example, as the poor cannot afford the outlay for larger boxes of it, or pay-as-you go mobile phone credit scratch cards. During the "Africa Rising" discourse in international business circles, it was felt that such markets might become primary drivers of revenue growth for TNCs. Indicative of this imaginative and empirical shift, Coca-Cola recently relocated its African headquarters from England to Johannesburg in South Africa (Grant 2015).

According to some urban theorists, cities in the Global South also serve as "holding centres" for populations that are surplus to the requirements of global capital accumulation (Davis 2006). However, even though slumification or slum growth is happening across much of the Global South, even these informal settlements are not just defined by exclusion from globalised economic networks. Even poorer cities and slums also serve productive functions (Grant and Nijman 2002), with some of the belts sold in Walmart, the world's largest retailer and private-sector employer (Muñoz *et al.* 2018), made in Dharavi, one of the biggest slums in Mumbai in India. Dharavi is reportedly the third-biggest slum in the world, with a population of approximately three-quarters of a million people in just over two square kilometres.

"Planning" and prefix urbanism in the Global South: smart, eco and other varieties

There have been a variety of approaches to improving conditions in slums. One approach is to try and upgrade the quality of public services provided in these retrospectively. Another approach has been so-called site and service schemes. This is where potential residents are provided with serviced plots of urban land and can then build their own dwellings. Other governments at times have engaged in slum clearance projects, although sometimes with political motivations. The most infamous of these is probably Operation Murambatsvina in Zimbabwe in 2005. Ostensibly this took place for planning reasons, although some argue that the government was trying to disrupt political opposition and reduce the size of the informal sector to allow government-affiliated companies to capture more market share.

We now live in a hyper-competitive world wrought by globalised capitalism and its accompanying philosophy – neoliberalism – and practice – neoliberalisation. The philosophy of neoliberalism was largely conceived, planned and rolled out in cities. Cities around the world compete with each other to be centres of innovation and investment and/or to carve out niches for themselves. Sometimes this competition may take somewhat surprising and seemingly progressive forms, such as Cape Town in South Africa marketing itself as a centre of gay tourism (Kanai 2014b). Cities around the world now engage in what is referred to as urban entrepreneurialism and try to assemble "growth coalitions" between the business and public sector to foster their competitive advantages in relation to other cities (Pillay 1994).

While urban entrepreneurialism may sound positive and it represents a competitive strategy in a neoliberalising urban environment, it can also generate new inequalities and exclusions by diverting investment and infrastructure away from other regions or districts. For example, in reference to Manaus – a city in the Brazilian Amazon – Kanai (2014a, 1080) notes:

> With the [new] bridge in operation, the old ferry service became obsolete overnight. The collapse of this activity particularly impacted its port base in Cacau Pirêra and surrounding communities where more than 18000 people reside. Media reports show that in addition to skyrocketing unemployment rates these areas face decreasing levels of public service and multiple insecurities, to which residents respond with intensifying protests – even attempting forceful interruptions to Manaus-bound vehicular traffic.

Thus increased connectivity between some places can result in disconnection for others. While we generally think improved infrastructure is a positive thing, there are opportunity costs associated with it, as money could be invested in other priorities and not everyone necessarily benefits. New infrastructure largely driven by demand from urban centres and planned from there may also generate environmental degradation. Ninety-five per cent of deforestation in the Amazon takes places within five kilometres of a road or navigable river (Barber *et al.* 2014) – another example of planetary urbanisation.

Flows of globalisation are often contradictory in terms of their developmental impact in cities. For example, flows of secondhand clothing into African cities may undercut local producers, at the same time as they allow people who might not otherwise be able to afford it to have access to clothing; although, ultimately, you cannot consume if you do not produce (at least at the level of a country or city). There are also many examples of more positive types of flows associated with globalisation. Flows of FDI in SEZs in China might be considered a more positive type of flow, although not, of course, for workers who may have lost their jobs in the Global North as a consequence of this, and the quality of work created is low.

There are also examples of progressive infrastructure planning of cities within the Global South. Perhaps one of the most famous of these is the cable car system linking the *favelas* (shanty towns) of Medellin in Colombia to the central business district of that city (see Figure 10.1).

This application of ski slope technology opened up employment opportunities in the central city to shanty town dwellers, although complementary investments in social services, such as education, were also important in improving socio-economic conditions (Brand and Davila 2011).

FIGURE 10.1 Cable car system in Medellin, Colombia

One of the cities which has attracted a lot of attention and praise in the literature is Curitiba in Brazil, which introduced dedicated bus lanes and made this form of transportation relatively accessible in the early 1990s. Addis Ababa, the capital city of Ethiopia, recently became the first city in sub-Saharan Africa to have a light rail metro system. Thus, there are many examples of progressive planning taking place across cities in the Global South. However, there are also counter-examples.

Sometimes in order to attract global investment city planning authorities set up what Martin Murray (2015) refers to as "city doubles". This type of urban regeneration entails constructing entirely new cities adjacent to old ones rather than renovating them. Another way of thinking about these new development initiatives is as what Michel Foucault (1984, 3) called heterotopias, or "a kind of effectively enacted utopia" (see Carmody and Owusu 2016). By trying to do away with some of the perceived disadvantages of the main city, such as high levels of crime in Johannesburg in South Africa through the construction of Waterfalls City outside of the city, the city double is an attempt to attract (foreign) investment by presenting a "clean slate".

Waterfall City has an estimated price tag of US$5.8 billion – the largest property development ever undertaken in Africa (Murray 2015a). This represents not just sub-urbanisation but also ex-urbanisation, where a new city is meant to be created *de novo*. It therefore represents an attempted erasure of the socio-historical conditions which have produced Johannesburg as a particular type of space. We can think of this strategy as one of connection through erasure. There are many examples of these types of proposed urban developments in Africa and other developing world regions. In Africa, examples include Eko Atlantic, which is built on land reclaimed from the Atlantic and was launched by former US president Bill Clinton.

Many of the proposed new cities also state that they are going to be ecologically friendly and/or smart. Smart cities are meant to use computerised technology, such as self-driving cars which can be more easily shared to reduce carbon emissions. Often these new cities are only "alleged eco-cities" driven by the usual imperatives of economic growth and political regime maintenance, such as Masdar City in Abu Dhabi (Cugurullo 2016). This is one aspect of these new "urban fantasies" (Watson 2014), many of which look like they are unlikely to move beyond the planning stage. For example, the Modderfontein development in South Africa, which was being marketed as the "New York of Africa" by its Hong Kong–based developer, recently collapsed.

These new types of developments serve not only economic but also ideological functions. "Worlding" practices, such as the construction of spectacular skyscrapers, serve to distract attention from new exclusions produced by urban economic restructuring (Roy and Ong 2011). Even slums can be repurposed to become spectacles through tourism, for example (Jones and Sanyal 2015).

Slum tourism is a highly controversial practice. On the positive side, it can bring income and economic activity to slums. On the other hand, it is quite a voyeuristic practice – touring through poverty as a form of entertainment or education, or perhaps what some have called edutainment. While researching a project on the impacts of ICT on tourism in South Africa, I took a "slum tour" in Cape Town in 2012. I did it because I wanted to research the different tourism products that were on offer in the city, but it was a deeply uncomfortable experience. We were picked up by the tour operator in their (old) Mercedes Benz and driven to hostels where there were multiple families living in one room and to see inside of shacks. While it was educational, I had a deeply ambivalent feeling about the morality of the exercise. I justified it on the basis that (some) money was being brought into those communities by this type of tourism.

The new urban agenda and generative urbanism?

As noted earlier, the SDGs were adopted in 2015. Goal 11 is Sustainable Cities and Communities. Following on from this the United Nations Conference on Housing and Sustainable Urban Development held a meeting in Quito, Ecuador, in 2016 where the "New Urban Agenda" (NUA) was adopted. This was subsequently brought to the United Nations General Assembly, where it was ratified (Schindler 2017).

Some regions of the developing world, such as most of sub-Saharan Africa, have experienced urbanisation without industrialisation. As noted earlier, this has been very problematic because of the lack of employment opportunities for people in cities and the consequent lack of taxes to fund urban infrastructure and housing. Some refer to this phenomenon as "over-urbanization" (Beall and Fox 2009). However, there are also many other cases in the Global South where industrialisation has not brought wide enough or distributed economic benefits to the majority of the population. This may even be the case when urbanisation is accompanied by industrialisation. If industrialisation is based on relatively simple or low-tech manufacturing, it is often foreign investors who benefit most through profit repatriation, and workers may be exposed to extreme risks, as employers may seek to maximise profits by having production take place in sub-standard or unsafe factories. There have been several clothing factory collapses in Bangladesh, one of which killed more than 1,000 people in 2013 – the Sarvar building collapse.

The NUA talks of the need for better urban governance and integrated planning – things that nobody, or very few people, could be against. However, a fundamental problem is that many cities in the Global South remain largely parasitic rather than generative (Hoselitz 1955). Generative cities create wide-scale employment and taxes, whereas parasitic cities' economies are largely based on import-export trade and elite consumption.

China would be amongst the primary examples of generative urbanism in recent decades. While there have been examples of "ghost cities" being built in China, the scale, pace and generally generative nature of urbanism in that country have been striking. Shenzhen Municipality, which is directly adjacent to Hong Kong, is a case in point. There the population of Baoan County grew from 30,000 in 1980 to 7 million by 2000 (McGee *et al.* 2007). China's development, particularly in recent years, has also been characterised by substantial technological up-grading (Zhou 2007) and dramatic reductions in poverty. As hundreds of millions of people have been lifted out of poverty in China, there are now "only" an estimated 43 million people living in poverty in that

FIGURE 10.2 Shenzhen, China

country, and the government there has devised personalised poverty plans to eliminate this by 2020 (Economist 2017).

As the planet urbanises, urban development will become increasingly synonymous with development more broadly. Finding ways to make cities generative and sustainable is probably the key development challenge of the future. Currently, however, most of the world's poor are found in rural areas, and these are being increasingly affected by the changing climate. The next chapter examines rural development and the implications of climate change.

Note

1 Some of this section draws on ideas from Carmody and Owusu (2016).

Further reading

Books

Chant, S., and C. McIlwaine. 2016. *Cities, Slums and Gender in the Global South: An Anatomy of a Feminised Urban Future*. London: Routledge.

Gilbert, A., and J. Gugler. 1992. *Cities, Poverty and Development: Urbanisation in the Third World*. Oxford and New York: Oxford University Press.

Articles

Malera, J., R. Grant, M. Oteng-Ababio, and B. Ayele. 2013. "Downgrading – An Overlooked Reality in African Cities: Reflections on an Indigenous Neighborhood in Accra." *Ghana Applied Geography* 36 (1): 23–30.

Satterthwaite, D., G. McGranahan, and C. Tacoli. 2010. "Urbanization and Its Implications for Food and Farming." *Philosophical Transactions of the Royal Society B-Biological Sciences* 365 (1554): 2809–20.

Websites

African Centre for Cities, www.africancentreforcities.net

C40 Cities Climate Leadership Group, www.c40.org

UN-Habitat, The United Nations Human Settlements Programme, https://unhabitat.org/

References

Afenah, A. 2009. "Conceptualizing the Effects of Neoliberal Urban Policies on Housing Rights: An Analysis of the Attempted Unlawful Forced Eviction of an Informal Settlement in Accra, Ghana." *Development Planning Unit.* University College, London. www.bartlett.ucl.ac.uk/dpu/latest/publications/dpu-working-papers/WP139_Afia_Afenah_Internet_copy.pdf.

Appadurai, A. 1995. "The Production of Locality." In *Counterworks: Managing the Diversity of Knowledge*, ed. R. Fardon. Abingdon and New York: Routledge.

Barber, C.P., M.A. Cochrane, C.M. Souza, and W.F. Laurance. 2014. "Roads, Deforestation, and the Mitigating Effect of Protected Areas in the Amazon." *Biological Conservation* 177: 203–9.

Barkawi, T. 2006. *Globalization and War.* Lanham, MD: Rowman and Littlefield.

Beall, J., and S. Fox. 2009. *Cities and Development.* New York: Routledge.

Boone, C. 2014. *Property and Political Order in Africa: Land Rights and the Structure of Politics.* New York: Cambridge University Press.

Boserup, E. 1993. *The Conditions of Agricultural Growth: The Economics of Agrarian Change Under Population Pressure.* London: Routledge.

Brand, P., and J. Davila. 2011. "Aerial Cable-Car Systems for Public Transport in Low-Income Urban Areas: Lessons From Medellin, Colombia." In *World Planning Schools Congress: Planning's Future – Futures Planning: Planning in an Era of Global (Un)certainty and Transformation.* Perth, Australia, July 4th.

Brenner, N. 2015. "Introduction: Urban Theory Without an Outside." In *Implosions/Explosions: Towards a Study of Planetary Urbanization*, ed. N. Brenner. Berlin: Jovis.

———., and C. Schmid. 2015. "Towards a New Epistemology of the Urban." *City* 19 (2–3): 151–82.

Bryceson, D.F., and T. Mbara. 2003. "Petrol Pumps and Economic Slumps: Rural-Urban Linkages in Zimbabwe's Globalisation Process." *Tijdschrift Voor Economische En Sociale Geografie* 94 (3): 335–49.

Carmody, P., and D. Taylor. 2016. "Globalisation, Land Grabbing and the Present-Day Colonial State in Uganda: Ecolonisation and Its Impacts." *Journal of Environment and Development* 25 (1): 100–26.

———., and F. Owusu. 2016. "Neoliberalism, Urbanization and Change in Africa: The Political Economy of Heterotopias." *Journal of African Development* 18 (1): 61–73.

Castells, M. 1994. *Distinguished Visiting Lecturer.* Minnesota: MacAlester College.

———., and A. Portes. 1989. "World Underneath: The Origins, Dynamics, and Effects of the Informal Economy." In *The Informal Economy: Studies in Advanced and Less Developed Countries*, eds. A. Portes, M. Castells and L.A. Menton. Baltimore, MD: John Hopkins University Press, pp. 11–41.

Cheeseman, N. 2018. "Will Urban Innovation Solve Africa's Development Challenges?" Keynote address at *Development Studies Association of Ireland Annual Conference.* Dublin, October.

Cugurullo, F. 2016. "Urban Eco-Modernisation and the Policy Context of New Eco-City Projects: Where Masdar City Fails and Why." *Urban Studies* 53 (11): 2417–33.

Davis, M. 2006. *Planet of Slums*. London: Verso.
The Economist. 2017. "The Last, Toughest Mile: China's New Approach to Beating Poverty." www.economist.com/news/china/21721393-after-decades-success-things-are-getting-harder-chinas-new-approach-beating-poverty.
Foucault, M. 1984. *Of Other Spaces: Utopias and Heterotopias*. http://web.mit.edu/allanmc/www/foucault1.pdf.
Giddens, A. 1990. *The Consequences of Modernity*. Stanford, CA: Stanford University Press.
Grant, R. 2015. *Africa: Geographies of Change*. Oxford: Oxford University Press.
———., and J. Nijman. 2002. "Globalization and the Corporate Geography of Cities in the Less Developed World." *Annals of the Association of American Geographers* 92 (2): 320–40.
Hampwaye, G., E. Nel, and C.M. Rogerson. 2007. "Urban Agriculture as Local Initiative in Lusaka, Zambia." *Environment and Planning C-Government and Policy* 25 (4): 553–72.
Hoselitz, B.F. 1955. "Generative and Parasitic Cities." *Economic Development and Cultural Change* 3 (3): 278–94.
Jones, G.A., and R. Sanyal. 2015. "Spectacle and Suffering: The Mumbai Slum as a Worlded Space." *Geoforum* 65: 431–9.
Kanai, J.M. 2014a. "On the Peripheries of Planetary Urbanization: Globalizing Manaus and its Expanding Impact." *Environment and Planning D-Society & Space* 32 (6): 1071–87.
———. 2014b. "Whither Queer World Cities? Homo-Entrepreneurialism and Beyond." *Geoforum* 56: 1–5.
King, A. 1990. *Urbanism, Colonialism and the World-Economy: Cultural and Spatial Foundations of the World Urban System*. London: Routledge, 1991.
Klein, N. 2014. *This Changes Everything: Capitalism Vs. The Climate*. London: Allen Lane.
Li, T. 2007. *The Will to Improve: Governmentality, Development, and the Practice of Politics*. Durham, NC and London: Duke University Press; Chesham: Combined Academic [distributor].
Malaquais, D. 2007. "Douala/Johannesburg/New York: Cityscapes Imagined." In *Cities in Contemporary Africa*, eds. M.J. Murray and G.A. Myers. Basingstoke: Palgrave Macmillan, pp. 31–52.
McGee, T.G., G. Lin, A. Marton, M.C.H. Wang, and J. Wu. 2007. *China's Urban Space: Development Under Market Socialism*. London: Routledge.
Muñoz, C., B. Kenny, and A. Stecher. 2018. *Walmart in the Global South: Workplace Culture, Labor Politics, and Supply Chains*. Austin: University of Texas Press.
Murray, M.J. 2015. "Waterfall City (Johannesburg): Privatized Urbanism in Extremis." *Environment and Planning A* 47 (3): 503–20.
———., and P. Carmody. 2015. *Africa's Information Revolution: Technical Regimes and Production Networks in South Africa and Tanzania*. Chichester, West Sussex, UK and Malden, MA: John Wiley & Sons Inc.
Mustapha, A. 1992. "Structural Adjustment and Multiple Modes of Livelihood in Nigeria." In *Authoritarianism, Democracy and Adjustment: The Politics of Economic Reform in Africa*, eds. P. Gibbon, Y. Bangura, and A. Ofstad. Uddevalla, Sweden: Scandinavian Institute for African Studies.
Myers, G. 2018. "The Africa Problem in Global Urban Theory: Reconceptualising Planetary Urbanization." *International Development Planning Review* 10, 231–253.
Peck, J., and A. Tickell. 2002. "Neoliberalizing Space." *Antipode* 34: 380–404.
Pillay, U. 1994. "Local Government Restructuring, Growth Coalitions, and the Development Process in the Durban Functional Region, c. 1984–1994." *Urban Forum* 5 (2): 69–86.
Portes, A., and J. Walton. 1981. *Labor, Class and International System*. New York: Academic Press.
Potts, D. 2010. *Circular Migration in Zimbabwe & Contemporary Sub-Saharan Africa*. Woodbridge: James Currey.
Robinson, J. 2006. *Ordinary Cities: Between Modernity and Development*. London: Routledge.
Roy, A., and A. Ong. 2011. *Worlding Cities: Asian Experiments and the Art of Being Global*. Chichester: Wiley-Blackwell.
Santos, M. 1979. *The Shared Space: The Two Circuits of the Urban Economy in Underdeveloped Countries*. London: Methuen.
Schindler, S. 2017. "The New Urban Agenda in an Era of Unprecedented Global Challenges." *International Development Planning Review* 39 (4).
Selwyn, B. 2014. *The Global Development Crisis*. Cambridge: Polity.

———. 2017. *The Struggle for Development*. Cambridge and Malden, MA: Polity.

Simone, A.M. 2004. *For the City Yet to Come: Changing African Life in Four Cities*. Durham, NC: Duke University Press.

Storper, M., and A.J. Scott. 2016. "Current Debates in Urban Theory: A Critical Assessment." *Urban Studies* 53 (6): 1114–36.

Surborg, B. 2011. "World Cities are Just Basing Points for Capital: Interacting with the World City from the Global South." *Urban Forum* 22 (4): 315–30.

United Nations, Department of Economic and Social Affairs, Population Division. 2014. *World Urbanization Prospects: The 2014 Revision, Highlights. (ST/ESA/SER.A/352)*. http://esa.un.org/unpd/wup/Highlights/WUP2014-Highlights.pdf.

UN-Habitat. 2008. *The State of African Cities 2008: A Framework for Addressing Urban Challenges in Africa*. Nairobi: UNHabitat.

———. 2013. *State of the World's Cities 2012–13: Prosperity of Cities*. Nairobi: UN-Habitat.

Watson, V. 2014. "African Urban Fantasies: Dreams or Nightmares?" *Environment and Urbanization* 26 (1): 215–31.

World Bank. 2009. *World Development Report 2009: Reshaping Economic Geography*. Washington, DC: World Bank and London: Eurospan [distributor].

Zhou, Y. 2007. *The Inside Story of China's High-Tech Industry: Making Silicon Valley in Beijing*. Lanham: Rowman and Littlefield.

CHAPTER

11

Rural development and climate

Crisis and transcendence?

The global capitalist economy is marked by a number of tendencies and dynamics. It is based on growth as firms seek to expand the amount they produce and thereby garner higher profits. "More is more" could be seen to be the mantra of business. However, this growth dynamic has a variety of severe, and from a global social and environmental perspective, highly undesirable consequences ranging in scale from climate change (global), to intense deforestation in particular localities (which may interact to increase their overall impact), to slumification. For example, when trees are burnt their carbon capture or sequestration function is gone and carbon dioxide is released into the atmosphere, which is a major cause of climate restructuring. The term climate restructuring is preferred to climate change or even climate disruption (Maass 2009) for a number of reasons. Climate as a term implies regularity, in contrast to "change". Ultimately humans are dependent on the "natural" environment for their survival.

The climate crisis

There are three types of natural resources – (re)source, sink and service (Grafakos *et al.* 2016). Source resources provide inputs into production, such as timber to make tables. Sink resources absorb the by-products of production, such as the oceans or atmosphere absorbing carbon dioxide. Service resources are things that people need to survive, such as air or water. The global market economy tends to degrade all three types of these resources through overexploitation associated with the profitability principle. As global economic output expands, more and more resources get used and their quality globally degrades, as the climate system is disrupted and produces more unpredictable weather, for example. The contradiction between the growth dynamic of the global economy and relatively fixed endowment of natural resources is what O'Connor (1991) calls "the ecological contradiction of capitalism".

In 2015 the Paris Climate Change Agreement was signed. This was meant to offer a roadmap to prevent the world from breaching the "planetary boundary" (Steffen *et al.* 2015) around climate, when there would be catastrophic and irreversible change characterised by what systems scientists call deviation amplification or "positive feedback". For example, it is thought that the Siberian tundra contains 70 billion tonnes of methane. "If even a small fraction of this escapes, it will eclipse the estimated 600 million metric tons of methane that are emitted each year, from natural and human sources, and cause a dramatic acceleration in global warming" (Toulmin 2009, 4). Furthermore, the more the earth's climate heats up, the more CO_2 plants and soils release – a mechanism known as carbon feedback, which could help increase the average temperature on earth by 6 degrees Celsius by 2100 (Rockstrom *et al.* 2009 cited in Parr 2013).

There are a number of problems with the Paris Agreement, however. One of these is effectiveness as the Trump administration in the US announced it would pull out of it in 2017, although this takes a number of years to come into force. The United States is the world's second-biggest greenhouse gas emitter after China. A second problem is that the agreement is based on voluntary "nationally determined contributions". There is no enforcement mechanism beyond the moral suasion and "peer pressure" of other states who signed the agreement and other interested actors.

The science behind climate change modelling is technical, allowing some climate change deniers to put forward their position with a force amongst the general public that it might not otherwise have. However, there is scientific consensus, as expressed through the UN Inter-Governmental Panel on Climate Change's (IPCC) periodic reports that people-made, or anthropogenic, climate change is happening and presents severe dangers. The dangers, however, vary geographically.

The earth could sustainably absorb about two tonnes of carbon dioxide per year from everyone on the planet. However, there are vast disparities in the levels of greenhouse gas emissions per capita. The average carbon dioxide emissions from someone in Burundi is 800 times lower than that of someone in the United States, for example (calculated from World Bank 2014). While South Africa is a major greenhouse gas emitter, most other African economies are "green", on this measure, while many developed countries exceed the two-tonne limit by up to ten times. For example, largely because of its agricultural sector, Ireland produces an average of 13 tonnes of carbon dioxide equivalent per person per year (calculated from Ó Fátharta 2016), having reduced these by 17 per cent from 2005. By way of comparison, someone in Burundi produces an average of 0.1 tonne per person (World Bank 2014), whereas the average for Africa as a whole is 1 tonne. However, because Burundi and most other African countries' populations depend on rain-fed agriculture to support most of their populations, they are likely to be the worst affected by climate change. This represents a form of climate injustice, as those who have done least to create the problem will be the worst affected by its impacts. For example, glaciers in the Andes have lost 20 per cent of their volume since 1970, but they provide a critical source of water supply for Bolivia, where it is reported that domestic water taps have dried up in El Alto (Parr 2013).

While Africa is likely to be the continent most affected by climate change, some governments there have been making progress in terms of sustainable development. In contrast to common Western images of poverty-driven environmental despoliation in Africa,

> in a study measuring the long-term sustainable development of thirty-one countries worldwide by Standard Chartered Bank in 2013, Ghana, Uganda, Egypt, and

> Nigeria all came in the top ten for progress over 2000–2102 in an array of sustainable development indicators. On Yale University's Environmental Performance Index (EPI), which ranked those states which have improved most over a ten-year period in terms of indicators on human health and protection of ecosystems, in 2014 Niger came in first, Sierra Leone fifth, Namibia sixth and Congo seventh.
> (Death 2016, 7)

Although such gains could be wiped out by climate restructuring. For example, over half of all Ugandan exports are coffee, but according to that country's Department of Meteorology just a small change in the temperature could wipe out most of that country's coffee crop (Toulmin 2009). A Climate Institute (2016) study found that climate change could reduce the area suitable for coffee cultivation worldwide by 50 per cent, which would spark both substantial hardship for many communities and massive price rises.

The IPCC estimates that the value of African crops could fall by up to 90 per cent by the turn of the century as a result of anthropogenically induced climate change (Toulmin 2009). There are many other negative impacts of climate restructuring in Africa, ranging from coastal inundation to flooding and droughts, depending on the region of the continent under consideration. African delegates at UN climate summits have begun to use words such as genocide to describe the impacts of climate change on the continent (Klein 2014), although perhaps somewhat ironically, given its human rights' history and the fact it is an oil exporter, it was the Sudanese representative at one of these conferences that used this terminology.

The impacts on other parts of the Global South will also be extremely severe. For example, the US National Intelligence Director (quoted in Parenti 2011, 139) argued

> For India, our research indicates the practical effects of climate change will be manageable by New Delhi through 2030. Beyond 2030, India's ability to cope will be reduced by declining agricultural productivity, decreasing water supplies, and increasing pressures from cross-border migration into the country.

Of course, the US security apparatus has an interest in exaggerating threats in order to secure funding. The US Department of Defense wrote a report in 2004 in which it claimed that Britain could have a "Siberian" climate in less than 20 years (cited in Townsend and Harris 2004). Large parts of eastern China – one of the most populous parts and biggest food-producing regions of the country – look likely to become uninhabitable. One study led by the Massachusetts Institute of Technology found that by the end of the century in eastern China the "wet bulb temperature" (which combines heat and humidity) might periodically rise over the 35-degree-Celcius threshold. At that point "the air is so hot and humid that the human body cannot cool itself by sweating and even fit people sitting in the shade die within six hours" (Carrington 2018).

Climate change is also increasingly being securitised. "A security problem arises when someone – a person, gang or group, or state – threatens another's life, limb or livelihood" (Kolodziej 2005, 1). However, it is not just people which now are seen to present a security threat but also global environmental change. Some argue security is an inherently conservative concept, as it is about keeping things as they are, even if they are unjust (Hayes 2016), although this is not the case when there is widespread conflict and attempts to establish order. In any event, it does not appear that this will be possible in the context of global climate restructuring. According to the head of the International Consultative Group for International Agricultural Research (CGIAR):

> The annual production gains [in food production] we have come to expect will be taken away by climate change. We are not so worried about the total amount of food produced so much as the vulnerability of the one billion people who are without food already and who will be hardest hit by climate change. They have no capacity to adapt.
>
> (Quoted in Vidal 2013)

While this may be overstated, as peasant farmers have shown great resilience and adaptability to often difficult conditions, global warming will create severe new challenges for them. Crop yields are likely to be most negatively affected in tropical areas, many of which already have high incidences of hunger (Wheeler and von Braun 2013 cited in Buxton *et al.* 2016).

Although much international attention has been given to the concept of climate refugees, people need to have resources or money to migrate, so this "threat" may be exaggerated (Raleigh *et al.* n.d.). However, international migration is also being increasingly securitised, as evidenced by Donald Trump's proposed border wall with Mexico, for example, or through European Union support for Libya to securitise its southern border to try to prevent migrants from sub-Saharan Africa from making the crossing over the Mediterranean. Securitising an issue, or casting it as a security threat, is a discursive move which allows for out-of-the-ordinary or emergency measures to be justified and taken. Paul Rogers has called this strategy liddism – that is, trying to keep a lid on problems by repressing dissent, or sometimes sponsoring militias and civil wars and securitising borders (cited in Hayes 2016). Securitisation also has other axes or dimensions, such as the repression of civil society. For example, in India in 2015 the government froze Greenpeace's bank account on the basis that the group was "anti-development" (Hayes 2016).

Reducing of emissions from deforestation and forest degradation or carbon colonialism?

In addition to the Paris Climate Change Agreement there have been other United Nations initiatives to try and offset the impacts of climate change. After the thirteenth meeting of the United Nation's Framework Convention on Climate Change (UN FCCC), a number of countries developed strategies to reduce emissions of carbon dioxide by increasing the protection of their existing forests and the planting of new ones (including plantations) in order to mitigate climate change (Murdiyarso *et al.* 2012 cited in Carmody and Taylor 2016). These strategies came together in what was to become known as the United Nations' Reductions of Emissions from Deforestation and Forest Degradation and enhance forest carbon stocks (REDD) programme, which subsequently became known as REDD+. The plus sign denotes purported benefit to local communities in addition to the programme goal of reducing the risks of dangerous climate change. Since it came into being, financial assistance has been made available from the World Bank and a variety of bilateral donors. This financial aid has created an incentive for some governments to protect forests by excluding locals and also to convert existing vegetation to plantations. In both cases, people have been either dispossessed of their land or denied access to resources they formerly used (Larson *et al.* 2013). In some cases, this has been associated with extreme violence.

In one carbon offsetting project in Uganda:

> Evictions led by UWA [Uganda Wildlife Authority] staff, police, and soldiers from the Uganda's People's Defense Force were violent to the extreme: people were killed, beaten and tortured, and women were gang-raped. In the process of resettling the Benet people, the UWA set fire to their homes, destroyed their crops, and confiscated cattle that remained inside the "red line" (the UWA boundary line of red sinking markers). When queried about the level of brutality waged against the indigenous population, one UWA official explained: "Mount Elgon National Park is an international conservation area. So we have to protect it from destruction".
>
> (Parr 2013, 32)

Land conversion often results in the release of carbon through forest clearance and burning, and new land uses are often far less effective in carbon sequestration. For example, according to one author of a report on the development of biofuel plantations "the clearance of grassland released 93 times the amount of greenhouse gases that would be saved by the fuel made annually on that land" (Fargione quoted in Shiva 2008, 80), thus actually accelerating the pace of climate change. Adrian Parr (2013, 11) argues that "capitalism appropriates limits to capital by placing them in the service of capital". So, for example, global reinsurance companies, which make much of their money by insuring smaller insurance companies, make billions in profits by selling developing countries, who have often done little to create the climate crisis, new protection or insurance schemes (Klein 2014). There are also questions about whether new plantations developed under the auspices of REDD+ will survive pressure for deforestation and changes in rainfall and temperature (Toulmin 2009).

While some developed countries have made substantial progress in the reduction of carbon dioxide emissions in recent years, this has partly been accomplished by the outsourcing of manufacturing to China (Toulmin 2009), rather than domestic production, thereby not contributing to an overall global reduction in greenhouse gas emissions. From 2002 to 2008 nearly half of China's total greenhouse gas emissions (the world's largest) related to producing for export markets (Malm 2012 cited in Klein 2014).

Climate change represents not just an existential threat to humanity over the long term but also in the shorter term to the idea and practice of development, as it has the potential to reverse the global gains that have been achieved in recent decades. It presents both an example of "the tragedy of the commons" and a major collective action or coordination problem (Okereke and Massaquoi 2017). The tragedy of the commons (Hardin 1968) is when a communal resource is overused because there is no cost attached to doing so, as has been the case historically for polluting the atmosphere. A collective action problem arises when individual and collective incentives are misaligned. For example, someone might think that there was little point in reducing their carbon dioxide emissions individually, whereas other people will not. Traditionally states have been the actors that societies have looked to solve collective action problems; however, as climate change is a global issue, states, too, suffer from the issue of coordination problems – hence the need for international agreements, such as the Paris Climate Change Agreement signed in 2015 in this area.

While China is now the world's largest greenhouse gas emitter, the rich or industrial countries are responsible for the majority of the stock of anthropogenic greenhouse gases in the atmosphere. There are thus issues around the geography of climate injustice and inter-temporal issues around this as well, as those who created the problem in the past may not have to pay its costs in the future. Climate change then presents itself as something which is fundamentally unjust in its impacts, as it is often the poor who have done the least to create the problem and have the fewest resources to cope with its impacts but are disproportionately paying for the historical overconsumption of fossil fuels by the rich world. In order to redress this there is a need for climate justice, which has been championed globally through the work of the former president of Ireland, Mary Robinson, through her Climate Justice Foundation.

Climate justice is the idea that there is a need to redress the imbalances of power which have created the current conjuncture. A number of principles underlie the concept:

- That those who have the greatest responsibility for the problem of climate change have the greatest responsibility to help resolve it.
- That those who can most afford it should contribute most to the solution.
- That economic, social and political rights should be enhanced by policy and legal frameworks to redress climate change and that they should not compromise the right to development. The UN Declaration on the Right to Development was proclaimed in 1986.

Climate justice will require a global transition to a low-carbon economy in addition to innovations in adaptation and mitigation.

As the SDGs recognise, there needs to be both a green transition and an elimination of poverty. There are some hopeful signs in relation to the possibilities of a green transition. A recent report by the Renewable Energy Policy Network found that for the first time annual investment in renewable energy projects had exceeded those in fossil fuels globally (McDonald 2016), although they still only account for about 13 per cent of total output. This is a major shift, although there are also issues such as reliability and the embedded energy in wind turbines, for example. James Lovelock, the world-famous environmentalist who developed the Gaia hypothesis that organic and inorganic entities on earth synergise to self-regulate, has warned that because wind is unreliable these turbines must be combined with power plants. Also significant energy is expended in the production of these turbines, which then has to be recouped (Crawford 2007).

Water scarcity

Globalisation is partly constituted through the largely unrestricted flow of commodities around the world, based on profitability criteria and demand. In some cases, this results in socially and environmentally perverse outcomes, even if these generate substantial profits for some actors. For example, it is well known that the Netherlands is a global centre for the international flower trade. It serves as a hub for the European flower trade in particular. Thus, cut flowers from Kenya, which is a country under acute water stress (Snyder n.d.), get exported to Europe. This represents a transfer of what is called virtual water in the form of the water that was used to grow the flowers (Allan 2011). It is important to remember though that emissions from flower production in Kenya and flying them to Britain can be under a fifth those produced by their equivalents being grown in lighted and heated greenhouses in the Netherlands (www.dfid.gov.uk cited in Toulmin 2009).

Water scarcity is also an increasingly urgent development issue. In common with food there is enough freshwater on earth to go around for everyone, but it is highly unevenly distributed. Water scarcity, then, is as much a social as a "natural" artefact, and Professor Larry Swatuk of the University of Waterloo also questions the way in which water scarcity is often constructed and conceived:

> The dominant 'freshwater availability' maps that one finds around the world turn on our accepting the fact that if a country has less than 1700 cubic metres per capita per year of water then it suffers serious water issues. However, this measures only blue water availability (i.e., groundwater recharge and runoff availability) relative to population. Yet, of the water we individually consume, about 80–90 per cent of that is in our food, most of which derives from rain-fed agriculture, which is green water – i.e. transpired rainfall. So the dominant scarcity indicator is in fact a fiction. If you cast the narrative not as one of scarcity due to population divided by freshwater availability, but one of resource capture by the few and ecological marginalization of the many, you can see that we have lots of water but, in Mehta's terms 'socially constructed the scarcity'.
>
> (Correspondence with author in Manahan 2016, 193)

There are competing uses for water, and this can sometimes cause conflict. At one point the former head of Coca-Cola said that he wanted to make Coke the most popular beverage on earth, ahead of water, and that he wanted to see people having taps in their kitchens which would dispense Coke. In India the pumping of water from the underground table to supply Pepsi plants has been very controversial, as 9 out of its 34 plants in the country are in regions designated as water-stressed (Hall and Lobina 2012 cited in Manahan 2016). In 2017 more than a million traders in India were boycotting Coke and Pepsi because they felt it was unsustainable to use 400 litres of water to produce 1 litre of these beverages (Doshi 2017).

In 1994 Pepsi was also given permission to start 60 restaurants in India (Kentucky Fried Chicken and Pizza Hut). However, according to Shiva (1999) the intensive breeding of livestock for such restaurants is also extremely water intensive, with about 190 gallons of water per day needed per animal, or about ten times the amount a normal family in India is meant to use. Climate disruption and water scarcity are two of the greatest threats associated with the globalising market economy. According to the CIA, globalisation will create

> an even wider gap between regional winners and losers than exists today. Its evolution will be rocky, marked by chronic volatility and a widening economic divide, deepening economic stagnation, political instability, and cultural alienation. It will foster political, ethnic, ideological and religious extremism; along with the violence that often accompanies it.
>
> (Global Trends 2015 quoted in Cavanagh *et al.* 2002, 30)

One of the ways in which governments in the Global South have often sought to utilise water more efficiently is through the construction of large-scale dams. These dams are often meant to provide both "free" electricity and water for irrigation for more intensive types of agriculture; however, they have often been very controversial for a variety of reasons, including population displacement.

Large dams are quintessential high-modernist development projects (Sneddon 2015). Their popularity around the world was partly driven by the perceived success of projects like the Tennessee Valley Authority project in the 1930s in the US, which succeeded in generating both employment and electricity during the Great Depression. However, in the Global South large dams are often built by Northern or Chinese companies, and consequently many of the benefits "leak" out of the local economy, with dam turbines often being manufactured abroad. In order to construct dams, local ecosystems must be flooded, and there is often extensive population displacement, which in some countries is forcible or achieved by simply flooding residents in what are to become dam reservoirs. In Sudan, European companies were recently involved in the construction of the Chinese-funded Merowe Dam, which involved around 50,000 people being displaced in addition to protestors being shot dead. In the United States there has been a trend of dam removal to try to restore pre-existing ecosystems.

Two of the most controversial dams in recent decades have been the Three Gorges Dam project in China and the Narmada Dam project in India, both of which were initially to receive funding from the World Bank. Both projects involved the construction of a series of dams and in the cases of the Three Gorges resulted in the displacement of 1.3 million people – the population of the region of the Irish capital, Dublin.

As mentioned previously, large dams typically produce hydro-electricity to the benefit often of industries and urban dwellers who can afford it, but to the detriment of those who are displaced. Rural dwellers in close proximity to these dams may also see increases in water-borne diseases such as schistosomiasis. Recent research found that nearly half a billion people around the world have an elevated risk of exposure to this disease as a result of dams blocking the migratory routes of river prawns, which eat the snails which host the disease (Sokolow *et al.* 2017). The spread of malaria may also increase as the parasite is carried by mosquitos who require water pools to breed. A recent paper on dams and malaria in sub-Saharan Africa found that

> in the absence of changes in other factors that affect transmission (e.g., socio-economic), the impact of dams on malaria in SSA will be significantly exacerbated

by climate change and increases in population. Areas without malaria transmission at present, which will transition to regions of unstable transmission, may be worst affected.

(Kibret *et al.* 2016)

Feeding and farming for the future sustainably?

Not being able to get access to enough food is perhaps the ultimate expression of powerlessness. While 2009 was the first year in human history that more than a billion people were classified as hungry, grain grown to produce fuel was enough to feed 330 million people that year (Smith 2010). In 2012, based on the number of calories that were required to support "normal" activities, one and a half billion people were hungry. As noted earlier, the numbers rise to over two and a half billion for those undertaking "intense" activity (Food and Agriculture Organization of the United Nations 2012). Food and agriculture are then development crucibles involving issues of power, life, health, water, environment and sustainability.

The globalisation of value chains both offers potential and creates new challenges for agriculture in the Global South as more powerful actors are able to leverage their positions to the detriment of those with less power. So-called non-traditional agricultural exports are now as important to developing countries as traditional exports, but while the value of retail coffee sales has more than doubled since the early 1990s, coffee-producing countries received less than half of the value of coffee exports that they had ten years previously (Osorio 2002). What constitutes a non-traditional export varies by context. For example, it would include rose exports from Kenya and Ethiopia, as these were not traditionally exported. However, globalisation also offers some opportunities to utilise consumer power to achieve progressive redistribution within chains. Furthermore, some governments have become more assertive, as the Ethiopian government's patenting victory over Starbucks, which allows for higher prices for producers, has recently shown (Richey and Ponte 2011). The Ethiopian government won the right to trademark the names Yirgacheffe, Harrar and Sidamo under the World Intellectual Property Organization, the regions where much of the coffee in the country is grown, thereby enabling producers to charge and receive higher prices. Coffee originated in Ethiopia.

Opportunities for backward and forward linkages to input suppliers and agro-processing can also help in the development of national systems of innovation in agriculture (Cramer 1999; Muchie *et al.* 2003). Achieving a "better deal" for small farmers and firms from the Global South in international markets does not always entail "moving up the value chain", but also the production of mass-produced agricultural commodities (Ponte and Ewert 2009).

New ICTs, such as mobile phones, may also enable farmers to achieve higher prices for their produce, particularly if they are not locked into exploitative forward contracts with traders where they must sell to them in exchange for loans or fertiliser, for example (Molony 2008). Burrell and Oreglia (2015) also found this was the case for fishermen on Lake Victoria in Uganda, who were also dependent on credit.

A green revolution?

African farmers have shown themselves to be resilient in the face of adverse events and shocks and to be responsive to price and other incentives. With the right conditions, major achievements have been recorded, such as the more than doubling of Ethiopia's production of cereals and grain legumes over a decade and a half (Sanchez *et al.* 2009). Effective institutional matrices and incentives can unleash farmers' capabilities to dramatically increase productivity and contribute to poverty and hunger reduction, as well as exports (Stein 2008). Recognising agro-ecological diversity is key to this scaling up of effective interventions, such as Malawi's input subsidy programme (Denning *et al.* 2009) to achieve a sustainable and gender-equitable green revolution (Negin *et al.* 2009). This programme was introduced after the famine of 2002 in Malawi and was successful in dramatically increasing farm yields; however, it subsequently deteriorated into corruption. There are ways in which the potential for corruption of these types of programmes can be reduced by directly delivering vouchers to farmers via their mobile phones, as in Nigeria, for example. It will also entail investing in research in semi-arid tropical crops, such as sorghum and chickpeas, and effective policy interventions which take advantage of synergies between states, markets and civil society actors (Tendler 1997).

There may also be space for the judicious use of new agricultural technologies, such as genetically modified crops which do not require expensive fertilisers and are resistant to climate change and pests. An example of this would include stress-tolerant rice for Africa and South Asia (STRASA), although as Parr (2013, 80) notes "there is nothing new or innovative about the rice varieties STRASA is trialling; the modern varieties it promotes are 'copies' of folk varieties that exist free for all. The STRASA work is outright theft, pure and simple". Furthermore, very limited yield improvements have been recorded with genetically modified drought-resistant crops. This is the case for Monsanto's DroughtGuard corn, even under moderate drought conditions, as it is thought it would increase productivity by approximately 1 per cent (Gurian-Sherman 2012). From 2008 to 2010 at least 261 patents were filed for "climate ready" crops, with nearly 80 per cent of these controlled by six agribusiness TNCs (Klein 2014).

Whether or not genetically modified crops should be developed and used is a highly controversial topic and depends perhaps on agro-ecological conditions. "Less than one-third of Africa's cropland is planted to improved seed varieties, compared to 82 percent of cropland in Asia, and average cereal yields are less than one-third as high as in Asia" (Borlaug and Carter 2008, vii). Increased crop yields on existing land can help reduce deforestation (Swaminathan 1994 cited Paarlberg 2008). Some African countries, such as Zambia and Zimbabwe, have at times refused food aid on the basis that it has been genetically modified and that there were concerns about its health impacts. However, many medicines which are commonly consumed in developed countries are genetically engineered without ill health effects (Paarlberg 2008). Like so many things in international development, different sets of policy interventions will deliver different sets of winners and losers.

Different studies find different impacts of the "Green Revolution" in Asia, which began in the 1960s. The Green Revolution increased crop yields through the dissemination of improved seed varieties, although these often required fertiliser and irrigation beyond the reach of many small farmers. One World Bank study found that the average real incomes of small farmers nearly doubled and those of the landless rose by more than that in southern India between 1973 and 1994. Presumably incomes rose for the landless as a result of greater labour demand. However, care should be taken with income statistics, particularly bearing in mind Jodha's paradox, which found some villagers in a study in India better off by being able to wear shoes, for example, even as their incomes had declined (Chambers 1995). Cullather (2010 quoted in Buxton *et al.* 2016, 179) argues that "the green revolution epicentres – Pakistan, India, Sri Lanka, Bangladesh, Mexico, the Philippines and Indonesia – are all among the most undernourished nations, each with higher rates of adult and childhood malnutrition and deficiency diseases . . . than most Sub-Saharan countries". The rate of malnutrition amongst children under five in India is double the average for sub-Saharan Africa according to World Bank data (Save the Children 2016).

Parr (2013) notes some of the benefits of these new genetically modified rice varieties, which can withstand flooding in West Bengal in India, with more and more farmers using them. She is right to ask, though, about patent ownership and whether farmers have to go into debt to pay for new machinery to effectively cultivate these varieties. There has been a rash of farmer suicides in India in recent decades (Shiva 2008; Patnaik 2008), largely as a result of increasing, and at times suffocating, indebtedness.

There are a variety of different agricultural techniques which are more sustainable than those associated with the corporate food regime and agro-industry. These include restorative and conservation agriculture. Restorative agriculture seeks to use natural processes to replenish nutrient stocks and soil depth after agro-ecologies have been overexploited. Restorative agricultural techniques, such as mulching, can restore soil depth and productivity (Bananuka *et al.* 2000).

Other approaches to raising small-farmer income: fair trade and land titling

There have been a variety of other approaches to try to raise incomes of small farmers, including "fair trade". The idea behind fair trade is that the actual producers should receive a higher proportion of the final price of the products they produce. Small banana farmers may only receive 1 to 2 per cent of the final value of their crop when it is sold in Western supermarkets, for example (Barratt Brown 1993). Fair trade products are typically somewhat more expensive as a greater share goes back to the farmers, perhaps on the order of 7 to 8 per cent of the final price. However, fair trade has been critiqued for not changing the overall external orientation of many rural economies in the Global South.

Another way in which some governments have sought to raise the incomes of small-scale farmers is through land titling. However, while some African governments have recently engaged in private land titling efforts, these have sometimes been shown to be ineffective in raising productivity and may increase gender inequality if men are the predominant recipients of land titles (Musembi 2007). Furthermore, research has shown that in some contexts productivity is growing more quickly under customary tenures, where land is collectively owned, than privately registered ones (Green 2008). "Pre-distribution" which increases access to assets for the poor before production, can have a significant impact on poverty reduction. A key issue is how initiatives like this should be financed.

This book has explored some of the main theories and practices of development. By necessity, as a result of space limitations, it has had to be selective in its coverage, but one of the issues which it has implicitly engaged with is whether the development enterprise is a flawed endeavour or not. The concluding chapter addresses this question more explicitly.

Further reading

Books

Adams, W. 2008. *Green Development: Environment and Sustainability in a Developing World*. 3rd edition. London: Routledge.

Parenti, C. 2011. *Tropic of Chaos: Climate Change and the New Geography of Violence*. New York: Nation Books.

Articles

Borras, S. Jr., R. Hall, I. Scoones, B. White, and W. Wolford. 2011. "Towards a Better Understanding of Global Land Grabbing: An Editorial Introduction." *Journal of Peasant Studies* 38 (2): 209–16.

Lavers, T., and F. Boamah. 2016. "The Impact of Agricultural Investments on State Capacity: A Comparative Analysis of Ethiopia and Ghana." *Geoforum* 72: 94–103.

Websites

International Fund for Agricultural Development, www.ifad.org
Land Matrix, https://landmatrix.org/en/
United Nations Food and Agriculture Organization, www.fao.org/home/en/

References

Allan, J.A. 2011. *Virtual Water: Tackling the Threat to Our Planet's Most Precious Resource*. London: I.B. Tauris.

Bananuka, J.A., P.R. Rubaihayo, and J. Zake. 2000. "Effect of Organic Mulches on Growth Yield Components and Yield of East African Highland Bananas." *ISHS Acta Horticulturae* 540.

Barratt Brown, M. 1993. *Fair Trade: Reform and Realities in the International Trading System*. London: Zed Books.

Borlaug, N., and J. Carter. 2008. "Foreword." In *Starved for Science: How Biotechnology Is Being Kept Out of Africa*, ed. R. L. Paarlberg. Cambridge, MA and London: Harvard University Press.

Burrell, J., and E. Oreglia. 2015. "The Myth of Market Price Information: Mobile Phones and the Application of Economic Knowledge in ICTD." *Economy and Society* 44 (2): 271–92.

Buxton, N., Z. Brent, and A. Shattuck. 2016. "Sowing Insecurity: Food and Agriculture in a Time of Climate Crisis." In *The Secure and the Dispossessed: How the Military and Corporations Are Shaping a Climate-Changed World*, eds. N. Buxton and B. Hayes. London: Pluto.

Carmody, P., and D. Taylor. 2016. "Globalisation, Land Grabbing and the Present-Day Colonial State in Uganda: Ecolonisation and Its Impacts." *Journal of Environment and Development* 25 (1): 100–26.

Carrington, D. 2018. "Unsurvivable heatwaves could strike heart of China by end of century". https://www.theguardian.com/environment/2018/jul/31/chinas-most-populous-area-could-be-uninhabitable-by-end-of-century.

Cavanagh, J., S. Anderson, D. Barker, *et al.* 2002. *Alternatives to Economic Globalization: A Better World Is Possible*. San Francisco, CA: Berrett-Koehler.

Chambers, R. 1995. "Poverty and Livelihoods – Whose Reality Counts." *Environment and Urbanization* 7 (1): 173–204.

Climate Institute. 2016. "A Brewing Storm: The Climate Change Risks to Coffee." http://fairtrade.com.au/~/media/fairtrade%20australasia/files/resources%20for%20pages%20-%20reports%20standards%20and%20policies/tci_a_brewing_storm_final_24082016_web.pdf.

Cramer, C. 1999. "Can Africa Industrialize by Processing Primary Commodities? The Case of Mozambican Cashew Nuts." *World Development* 27 (7): 1247–66.

Crawford, R.H. 2007. "Life-Cycle Energy Analysis of Wind Turbines – An Assessment of the Effect of Size on Energy Yield." *Energy and Sustainability* 105: 155–64.

Death, C. 2016. *The Green State in Africa*. New Haven and London: Yale University Press.

Denning, G., P. Kabambe, P. Sanchez, A. Malik, R. Flor, R. Harawa, P. Nkhoma, C. Zamba, C. Banda, C. Magombo, M. Keating, J. Wangila, and J. Sachs. 2009. "Input Subsidies to Improve Smallholder Maize Productivity in Malawi: Toward an African Green Revolution." *PLOS Biology* 7 (1): 2–10.

Doshi, V. 2017. "Indian Traders Boycott Coca-Cola for 'straining water resources'." *The Guardian*. www.theguardian.com/world/2017/mar/01/indian-traders-boycott-coca-cola-for-straining-water-resources.

Food and Agriculture Organization of the United Nations. 2012. *The State of Food Insecurity in the World: Economic Growth Is Necessary but Not Sufficient to Accelerate Reduction of Hunger and Malnutrition*. Rome: Food and Agriculture Organization of the United Nations.

Grafakos, S., A. Gianoli, and A. Tsatsou. 2016. "Towards the Development of an Integrated Sustainability and Resilience Benefits Assessment Framework of Urban Green Growth Interventions." *Sustainability* 8 (461): doi:10.3390/su8050461.

Green, D. 2008. *From Poverty to Power: How Active Citizens and Effective States can Change the World*. Oxford: Oxfam Publishing.

Gurian-Sherman, D. 2012. "High and Dry: Why Genetic Engineering Is Not Solving Agriculture's Drought Problem in a Thirsty World." www.ucsusa.org/food_and_agriculture/our-failing-food-system/genetic-engineering/high-and-dry.html#.Wc0EA1tSyUk: Union of Concerned Scientists.

Hardin, G. 1968. "The Tragedy of the Commons." *Science* 162 (3859): 1243–8.

Hayes, B. 2016. "Colonising the Future: Climate Change and International Security Strategies." In *The Secure and the Dispossessed: How the Military and Corporations Are Shaping a Climate-Changed World*, eds. B. Hayes and N. Buxton. London: Verso. www.theguardian.com/environment/2018/jul/31/chinas-most-populous-area-could-be-uninhabitable-by-end-of-century.

Kibret, S., J. Lautze, M. McCartney, L. Nhamo, and G.G. Wilson. 2016. "Malaria and Large Dams in Sub-Saharan Africa: Future Impacts in a Changing Climate." *Malaria Journal* 15: 448. https://doi.org/10.1186/s12936-016-1498-9.

Klein, N. 2014. *This Changes Everything: Capitalism Vs. The Climate*. London: Allen Lane.

Kolodziej, E. 2005. *Security and International Relations*. Cambridge and New York: Cambridge University Press.

Larson, A.M., M. Brockhaus, W.D. Sunderlin, A. Duchelle, A. Babon, T. Dokken, T.T. Pham, I.A.P. Resosudarmo, G. Selaya, A. Awono, and T.B. Huynh. 2013. "Land Tenure and REDD Plus: The Good, the Bad and the Ugly." *Global Environmental Change-Human and Policy Dimensions* 23 (3): 678–89.

Maass, P. 2009. *Crude World: The Violent Twilight of Oil*. London: Allen Lane.

Manahan, M. 2016. "In Deep Water: Confronting the Water and Climate Crises." In *The Secure and the Dispossessed: How the Military and Corporations Are Shaping a Climate-Changed World*, eds. N. Buxton and B. Hayes. London: Verso.

McDonald, B. 2016. "Investment in Renewable Energy Exploding – But Not in Canada." *CBC News*. www.cbc.ca/news/technology/renewable-energy-investment-1.3614477.

Molony, T. 2008. "Running Out of Credit: The Limitations of Mobile Telephony in a Tanzanian Agricultural Marketing System." *Journal of Modern African Studies* 46 (4): 637–58.

Muchie, M., P. Gammeltoft, and B. Lundvall, (eds.) 2003. *Putting Africa First: The Making of African Innovation Systems*. Aalborg, Denmark: Aalborg University Press.

Murdiyarso, D., M. Brockhaus, W.D. Sunderlin, and L. Verchot. 2012. "Some Lessons Learned from the First Generation Of REDD Activities." *Current Opinion in Environmental Sustainability* 4 (6): 678–85.

Musembi, C.N. 2007. "De Soto and Land Relations in Rural Africa: Breathing Life into Dead Theories About Property Rights." *Third World Quarterly* 28 (8): 1457–78.

Negin, J., R. Remans, S. Karuti, and J.C. Fanzo. 2009. "Integrating a Broader Notion of Food Security and Gender Empowerment into the African Green Revolution." *Food Security* 1 (3): 351–60.

Ó Fátharta, C. 2016. "Ireland's CO_2 Emissions Third Highest in EU." *Irish Examiner*, 23 November. https://www.irishexaminer.com/ireland/irelands-co2-emissions-third-highest-in-eu-431895.html.

O'Connor, A. 1991. "On the Two Contradictions of Capitalism." *Capitalism, Nature, Socialism* 2 (3): 107–9.

Okereke, C., and A. Massaquoi. 2017. "Climate Change, Environment and Development." In *Introduction to International Development: Approaches, Actors, Issues and Practice*, eds. P. Haslam, J. Schafer and P. Beaudet. Oxford and New York: Oxford University Press.

Osorio, N. 2002. "The Global Coffee Crisis: A Threat to Sustainable Development." In *Submission to the World Summit on Sustainable Development*. http://www.ico.org/documents/globalcrisise.pdf.

Paarlberg, R. 2008. *Starved for Science: How Biotechnology Is Being Kept out of Africa*. Cambridge, MA and London: Harvard University Press.

Parenti, C. 2011. *Tropic of Chaos: Climate Change and the New Geography of Violence*. New York: Nation Books.

Parr, A. 2013. *The Wrath of Capital: Neoliberalism and Climate Change Politics*. New York: Columbia University Press.

Patnaik, U. 2008. "Imperialism, Resources and Food Security with Reference to the Indian Experience." *Human Geography* 1 (1).

Ponte, S., and J. Ewert. 2009. "Which Way Is "Up" in Upgrading? Trajectories of Change in the Value Chain for South African Wine." *World Development* 37 (10): 1637–50.

Raleigh, C., L. Jordan, and I. Salehyan. n.d. "Assessing the Impact of Climate Change on Migration and Conflict." http://siteresources.worldbank.org/EXTSOCIALDEVELOPMENT/Resources/SDCCWorkingPaper_MigrationandConflict.pdf. Washington: World Bank.

Richey, L., and S. Ponte. 2011. *Brand Aid: Shopping Well to Save the World*. Minneapolis, MN: University of Minnesota Press.

Rockstrom, J., W. Steffen, K. Noone, A. Persson, F.S. Chapin, E.F. Lambin, T.M. Lenton, M. Scheffer, C. Folke, H.J. Schellnhuber, B. Nykvist, C.A. de Wit, T. Hughes, S. van der Leeuw, H. Rodhe, S. Sorlin, P.K. Snyder, R. Costanza, U. Svedin, M. Falkenmark, L. Karlberg, R.W. Corell, V.J. Fabry, J. Hansen, B. Walker, D. Liverman, K. Richardson, P. Crutzen, and J.A. Foley. 2009. "A Safe Operating Space for Humanity." *Nature* 461 (7263): 472–5.

Sanchez, P.A., G.L. Denning, and G. Nziguheba. 2009. "The African Green Revolution Moves Forward." *Food Security* 1 (1): 37–44.

Save the Children, India. 2016. "Malnutrition in India Statistics State Wise." www.savethechildren.in/articles/malnutrition-in-india-statistics-state-wise.

Shiva, V. 2008. *Soil Not Oil: Climate Change, Peak Oil, and Food Insecurity*. London: Zed Books.

Smith, J. 2010. *Biofuels and the Globalization of Risk: The Biggest Change in North-South Relationships Since Colonialism?* London and New York: Zed Books; New York: Distributed in the USA exclusively by Palgrave Macmillan.

Sneddon, C. 2015. *Concrete Revolution: Large Dams, Cold War Geopolitics, and the US Bureau of Reclamation*. Chicago, IL: University of Chicago Press.

Sokolow, S.H., I.J. Jones, M. Jocque, D. La, O. Cords, A. Knight, A. Lund, C.L. Wood, K.D. Lafferty, C.M. Hoover, P.A. Collender, J.V. Remais, D. Lopez-Carr, J. Fisk, A.M. Kuris, and G.A. De Leo. 2017. "Nearly 400 million people Are at Higher Risk of Schistosomiasis Because Dams Block the

Migration of Snail-Eating River Prawns." *Philosophical Transactions of the Royal Society B-Biological Sciences* 372 (1722).
Steffen, W., K. Richardson, J. Rockstrom, S. E. Cornell, I. Fetzer, E. M. Bennett, R. Biggs, S. R. Carpenter, W. de Vries, C. A. de Wit, C. Folke, D. Gerten, J. Heinke, G. M. Mace, L. M. Persson, V. Ramanathan, B. Reyers, and S. Sorlin. 2015. "Planetary Boundaries: Guiding Human Development on a Changing Planet." *Science* 347 (6223).
Stein, H. 2008. *Beyond the World Bank Agenda: An Institutional Approach to Development.* Chicago, IL: University of Chicago Press.
Tendler, J. 1997. *Good Government in the Tropics.* Baltimore, MD: Johns Hopkins University Press.
Toulmin, C. 2009. *Climate Change in Africa.* London: Zed Books.
Townsend, M., and P. Harris. 2004. "Now the Pentagon Tells Bush: Climate Change Will Destroy Us." *The Guardian.* www.theguardian.com/environment/2004/feb/22/usnews.theobserver.
Vidal, J. 2013. "Millions Face Starvation as World Warms, Say Scientists." *The Guardian.* www.theguardian.com/global-development/2013/apr/13/climate-change-millions-starvation-scientists.
Wheeler, T., and J. von Braun. 2013. "Climate Change Impacts on Global Food Security." *Science* 341 (6145): 508–13.
World Bank. 2014. "World Bank Open Data." http://data.worldbank.org.

CHAPTER

12

Getting to or after development?

Whereas in the past people have traditionally looked to states to solve collective action problems, globalisation and the widespread shift from government to governance as a modality of authority has troubled the notion of the state as an ontologically distinct actor, if it ever was. As authority has become more diffuse at one level under globalisation (the so-called retreat of the state) and more concentrated in another in the hands of TNCs, who are driven by the profitability imperative rather than the public good, collective action problems loom large, particularly in the context of growing ecological scarcity. This contradiction arguably also destabilises the very notion of development, as the old path of carbon-intensive industrialisation is increasingly generating environmental dysfunction, as we have moved to a situation where, as one European Union official has put it, "carbon kills", even as the EU itself is a major emitter of CO_2.

While the SDGs are problematic for some of the reasons outlined earlier, they also contain important insights and roadmaps for the future. One of these is the recognition that what happens in one part of the world affects others; that is, that development is combined and uneven, as Trotsky noted in the early part of the twentieth century. They also highlight the importance of reducing inequality (Goal 11) and by implication the uneven development that has characterised global development to date. The goals also highlight interdependence and interconnections between them. A point of debate is the tools to achieve them, with economic liberalisation seeming to be the default method favoured in the SDGs, even though this is associated with, and arguably responsible for, much of the increased inequality and environmental degradation which the world has witnessed in recent decades.

What constitutes development is arguably partly about values and what different people may consider to be worthwhile. Relatively new metrics, such as "Gross National Happiness" or "The Happy Planet Index", suggest less attention should be paid to traditional indicators of "development", such as increases in economic output and more to ecological and social sustainability. One question which looms large is how international development should be financed.

Financing international development and the reconfiguration of global geopolitics

In contrast to what is commonly assumed, Africa is a net creditor to the international system, rather than a net debtor (Boyce and Ndikumana 2001). When the costs of profit repatriation by TNCs, tax evasion and climate change are subtracted from inflows of investment, aid and loans, Africa suffers from capital outflow of US$58 billion a year (Sharples *et al.* 2016 cited in Brooks 2017). Moreover more than US$50 billion a year is likely lost from the continent in illicit financial outflows (Report of the High Level Panel on Illicit Financial Flows from Africa 2015), and if the loss of extracted natural resources is added to the debit column, the continent is getting substantially poorer (Bond 2006), rapid rates of economic growth in recent years notwithstanding. As the former president of South Africa, Thabo Mbeki, notes in his foreword to the report on illicit financial flows from Africa, plugging these could contribute substantially to achieving development on the continent. Thus, the challenge of development is partly about stopping extraction of capital from the continent, or Global South, more generally and enabling productive reinvestment.

Naomi Klein (2014) lists a variety of ways in which very substantial financing for climate change adaptation and mitigation could be raised. These could equally apply to development and include cutting military budgets globally, to instituting a global financial transactions tax, commonly called a Tobin tax after the economist who developed the idea, which could raise hundreds of billions of dollars globally each year. However, these are unlikely to come to pass, particularly with the election of Donald Trump in the United States, who has sought to dramatically cut back on international development assistance, although this has been resisted by the Congress in the US. Prior to his election campaign Trump (2013) had tweeted that "Every penny of the $7 billion going to Africa as per Obama will be stolen – corruption is rampant!"

In the United Kingdom the Conservative Secretary for International Development said that British aid had to work in that country's national interest and help promote bilateral trade (Brooks 2017). Indeed, in the wake of the Brexit vote there was discussion in Whitehall, where the British Foreign and Commonwealth Office is headquartered, of a so-called "Empire 2.0", where Britain would seek to revive or revivify trading links with former colonies. However, according to Olusoga (2017) "this newfound focus on the Commonwealth feels uncomfortably akin to recent divorcees looking up their former partners on Facebook; and being shocked to discover that they have got married, had kids and moved on".

Populism[1] in the US and UK may undermine international commitments to international development, although the UK now has to give 0.7 per cent of gross national income to ODA annually by law as of 2015. Financing for development then is perhaps a matter of both internal resource mobilisation and enhanced and increased ODA, without economic conditionalities. As noted earlier, the rise of emerging powers is reconfiguring global geopolitics and the prospects for development. Furthermore, in Britain, for example, there has also been a substantial anti-aid backlash after the Brexit vote, with some arguing that the sex abuse scandals over the behaviour of some Oxfam workers in Haiti after the 2010 earthquake there, for example, was being stoked by right-wing media.

A post-Western world after development?

According to some theorists we are now living in a post-Western world, where the centuries-long hegemony of the West in world affairs is coming to an end (Stuenkel 2016). According to Hans Rosling the PIN (personal identification number) of the world is 1114 – 1 billion people in each of the Americas, Africa and Europe and 4 billion in Asia (Rosling 2012). The rise of Asia, and particularly China, is now a long-term and established trend in the global political economy. This is having innumerable impacts around the world, ranging from increased demand for resources to Donald Trump's admiration of authoritarian BRICS leaders such as Xi Jinping in China and Vladimir Putin in Russia (Rachman 2016).

The precise consequences of this global shift in political and economic power are unknown, but there are a variety of indications as to what this is likely to mean. According to Stuenkel (2016, 10):

> Rather than directly confronting existing institutions, rising powers – led by China – are quietly crafting the initial building blocks of a so-called "parallel order" that will initially complement, and one day possibly challenge today's international institutions. This order is already in the making; it includes, among others, institutions such as the BRICS-led (sic) New Development Bank and the Asian Infrastructure Investment Bank (to complement the World Bank), Universal Credit Rating Group (to complement Moody's and S&P), China Union Pay (to complement MasterCard and Visa), CIPS [Cross Border Inter-Bank Payments System] (to complement SWIFT) [Society for Worldwide Inter-Bank Transactions] and the BRICS (to complement the G7).

There is substantial debate in the literature as to whether or not the BRICS are in fact attempting to craft a new world order or are merely being absorbed into the existing one. According to Professor Ian Taylor (2017) it is transnational capitalist hegemony which is globalising, rather than the reach of "emerging powers". Instead, they are merely being absorbed into it. The development context is consequently changing rapidly and dramatically. The rise of the emerging powers presents both opportunities and threats for less developed countries.

The rise of Donald Trump in the United States and the Brexit vote in the United Kingdom reflect a populist backlash against globalisation. We may now be entering a period in which globalisation slows, or at least slows down, in terms of certain axes or channels, while increasing in terms of others. For example, the Trump administration's trade policies in the United States are more protectionist than those of his predecessors. The director of the White House National Trade Council, the economist Peter Navarro, has written provocative books with titles such as *Death by China: Confronting the Dragon – A Global Call to Action* (Navarro and Autry 2011) and *The Coming China Wars: Where They Will Be Fought and How They Can Be Won* (Navarro 2007). By way of example the US imposed anti-dumping duties of between 96.81 and 162.24 per cent on Chinese aluminium foil in 2017 (Wood 2017). The Trump administration also threatened to cut aid to countries that voted against it at the UN in relation to its support for Jerusalem being recognised as the capital of Israel in 2017. However, rather than entering a post-globalisation world, as some have argued, we are entering a world of actually existing alter-globalisation, where the Trump administration will attempt to promote American exports, in addition to import-substitution domestically. The shift to additive manufacturing, such as three-dimensional printing, may also deepen tendencies towards on-shoring in the currently developed countries. For example, an AR 15 assault rifle can be 3D printed now (Whitwam 2014).

There are also movements towards localisation in much of the developing world – for example, around food sovereignty – even as the extent of South–South trade and investment deepens substantially. South–South trade overtook North–North trade in importance in the late 2000s (Dahi and Demir 2017). At the same time Internet interconnectivity continues to grow quickly around the world, with more than half of the world's population online in 2017 (Internet World Stats 2017), and global environmental change continues apace. All of this is to say that globalisation is a dialectical phenomenon: it is not unidirectional, but rather characterised by periods and spaces of advance and retreat, even as the overall direction is "forward" moving – carried in part by emerging powers. For example, at the 19th National Congress of the Communist Party of China President Xi Jinping argued that China should lead the world in economic, military, political and environmental issues. He also restated China's commitment to economic globalisation.

Future economic historians may look back at the period of the North Atlantic Financial Crisis as a period of hegemonic transition. In an ironic twist of fate, whereas China's "century of humiliation" was driven by Western powers imposing unfavourable trade deals on the country, it is now China which seeks to keep and open up further markets around the world. In its international economic relations, the Chinese government plays a two-level game – promoting external neoliberalisation to achieve resource, investment and market access through the WTO, for example, while adopting a 'flexeconomy' approach domestically and in bilateral relations, ostensibly without conditionality. The current world order is characterised by what Stuenkel (2016) calls "asymmetric bipolarity", where China is in the process of becoming the world's foremost economic power, whereas the United States retains its military predominance. What will this new configuration mean for industrial development?

Prospects for industrial development: escaping the matrix?

Matrix governance attempts to create a particular type of subjectivity and responsibilisation for market subjects, both citizens and states. States are meant to formulate and implement market-conforming policies, whereas consumer-citizens, particularly in developing-country contexts, are meant to engage in self-responsibilisation. According to the discourse of matrix governance, the conditions for this to be achieved are put in place by liberalised markets. However, in competition there are winners and losers. The extent to which people are permitted to lose is dependent on international divisions of power and the nature of national state–society formations. In underdeveloped countries without welfare systems, the extent of the losses are perhaps limited in monetary terms but are potentially catastrophic in terms of livelihoods and life chances. However, the political economy of resistance is complex. People working at the margins of survival in the informal sector are often too busy to be politically active. Consequently, resistance appears to be relatively low amongst what were once called the lumpen proletariat, or people who are now largely superfluous to the dynamics of globalised capital accumulation, despite efforts to tap into the "fortune at the bottom of the pyramid" (Prahalad 2009). Uneven development, marginalisation and exclusion interact, shape and constitute identities in complex ways.

Matrix governance attempts to combine hegemonic market discipline with legitimacy and coercive force where that is deemed necessary. As a power complex, it has been shown to be relatively robust given its different channels of operation and influence and the variable geometry of different types of power which can be brought to bear or into play. An alternative – following Gibson-Graham (1996) – would be to further delink from market logics and proactively construct alternative economies based on the localisation of production and non-wage labour forms of commodity production, to create parallel economies and systems of provisioning which displace those currently dominant. However, this has proven difficult, and to be effective it would be necessary to more thoroughly transform the dominant mode of production: capitalism.

In order to achieve such a transformation struggles over identity would be central. Consumer subjectivity would have to be challenged, as would the attempted atomisation or anomie which is introduced by advertising. There are many examples of such alternative economies; however, the commodity form tends to drive them to competitive extinction or niches given the global dominance of the law of value (Leyshon and Lees 2003). This term is used in different ways. I use it here to mean the organisation of society primarily along the lines of profitability criteria.

Escape from matrix governance, then, ultimately is about escape from the law of value. This is not to suggest a return to moribund, authoritarian and repressive state capitalisms of the past, but to think of new, more imaginative ways of organising economies, perhaps more modelled on those of successful cooperatives, such as Mont Dragon in Spain (Whyte and Whyte 1988), although the experience with "actually existing" or former socialist systems has not been positive. Another possibility would be to pursue green capitalist industrialisation. Are there alternative modes of capitalist governance which might be more humane, equitable and less environmentally destructive than the current matrix version?

The question of the appropriate scale at which to resist the depredations of neoliberalism is a difficult one. Some countries, such as Ethiopia, have been able to successfully implement heterodox economic policies and achieve remarkably sustained and dramatic economic growth with the beginnings of structural transformation (Oqubay 2015). However, this is still a labour-exploitative and resource-intensive type of growth. Ethiopia has had more latitude in designing its economic policies than many other developing countries because it is not yet a member of the WTO, and it is geo-strategically important and consequently continues to receive substantial inflows of Western ODA. What it does demonstrate, however, is that it is still possible for some low-income countries to implement heterodox economic policies successfully, even in a broader context of neoliberal globalisation.

Immanent, as opposed to imminent or more guided, development plays an important role in industrial development in the Global South. For example, the "Chinese dragon" economy "inhales air" (natural resources) and "breathes" out fire – low-priced manufactured goods which displace producers in other world regions (Kragelund and Carmody 2016). Given the multifold advantages which Chinese firms often have, such as active state support and soft financing, a still booming domestic market, etc., this makes it very difficult for firms without natural protection in Africa, for example, to compete. Natural protection refers to things which makes trading products of commodities over long distances difficult. For example, soft drinks tend not to be imported but to be bottled close to their final point of consumption because there is so much weight gain from water being added to other ingredients and bottling when it takes place. In work on the determinants of differential inter-firm performance, it was found that while the textile and clothing industry in Zimbabwe suffered a massive contraction on the heels of economic liberalisation, mining boomed and some niche protective clothing suppliers with natural protection were able to benefit from this (Carmody 2001). The intensely competitive and highly uneven nature of the global economy makes the establishment and expansion of exporting African service and manufacturing firms vital to study, particularly as there may be important policy implications and lessons for national governments, international development agencies and others involved in industrial development.

Successful, albeit to date niche, industrialisation in Africa is an outcome of capability development and multiple axes of strategic coupling, which themselves reduce information and other market failures in relation to regional assets, shape factors markets in more advantageous ways and create incentives and institutions which, in turn, promote capability development, innovation and exports (Lall 1992). Understanding how and why these assemblages are constructed will be an important theoretical and policy challenge in development studies in the years ahead.

Actor–network theory has been criticised for having a flat ontology, which eliminates a focus on power differentials in favour of examining network configurations. The principle of generalised symmetry, where actants are thought to have as much agency as actors proper, has also come under criticism. However, an emphasis on networks and social relations and mutual constitution does not necessarily obviate the need to examine power relations and differentials. For example, recent work by Carmody and Kragelund (2016) has argued that while the literature on African agency is salutary, it ignores fundamental power differentials between actors and their contexts, such as the non-human actor of the market, which fundamentally limits the incentives and potentials to engage in different forms of non–market-based governance. Is the rise of non-Western powers changing this context?

While Japan is often considered an honorary part of the West, given the restructuring of its institutions which took place after the Second World War and the fact that it is a liberal democracy, this not the case for what is now the world's second-biggest economy, China. As a result of its export-oriented economy China has, by far, the world's largest foreign exchange reserves at over $3 trillion, having previously been close to $4 trillion, before some of these were run down as a result of the Chinese economic slow-down. The Chinese state is now reportedly the biggest lender in sub-Saharan

Africa, ahead of the World Bank, and has a policy of agnosticism, or 'no questions asked', in its foreign relations, which enables it to deal with both more democratic and extremely repressive states. The growing influence of China has given rise to a new model of state-guided and perhaps authoritarian development, backed up by substantial economic resources in the Global South. However, there are also indications that China is being integrated into the Western-led transnational state. For example, whereas previously China disavowed the concept of aid this has now been integrated into official discourse and there is evidence of convergence in the international aid architecture and regime (Kragelund 2014). While China is not a neoliberal state internally, it largely promotes neoliberalisation overseas to promote market and resource access (Carmody and Kragelund 2016). Does this new configuration mean the "end of development" as has been suggested?

The future or end of development?

Even as the "Age of Sustainable Development" appears to be at its apogee or zenith in the form of the SDGs, it is also perhaps a time of crisis for the idea of development for a number of reasons. The first of these is the ongoing global environmental crisis. According to the World Wildlife Fund there are fewer than half the number of vertebrates on earth now as there were 40 years ago, although there is some dispute over the methodology which was used in the 2014 Living Planet Report.

> Latin America, home to many low-income countries, has borne the brunt of animal loss, with 'a dramatic decline' of 83 percent cited in the report. The opposite was found in countries where income is high – those nations show a 10 percent increase in biodiversity.
>
> (Dell'Amore 2014)

However, there is also great potential for sustainable development. Solar radiation coming into the earth–atmosphere system represents about 10,000 times the commercial use of energy on the planet (Sachs 2008). Humans appropriate up to 50 per cent of the earth's net plant product and also its photosynthetic potential through land clearance (Sachs 2008). If this appropriation of net plant product continues to accelerate, this has further serious implications for biodiversity.

This book has reviewed a variety of different theories and practices of development, although there are many others which have not been covered. While there are differences between modernisation theory and neoliberalism and related approaches, they share an emphasis on the internal reasons for underdevelopment: supposed backward cultural practices, poor governance, inappropriate policies and corruption. However, as neoliberal policies have been applied across much of the Global South for the last four decades, other orthodox explanations were found for continued development failures. These include the theory of non-implementation or half-hearted implementation of economic reforms (Van de Walle 2001) and the theory of different types of traps such as poverty or conflict traps (Collier 2007), which are often theorised as being internally constituted by domestic social forces. As this book has argued, these are transnationally constituted, suggesting the need to reform global governance structures.

While these traps exist, what is often left undiscussed is the ways in which they have developed historically and are replicated through interactions with other places. For example, as discussed earlier, the role which resources sometimes play in providing opportunities for and helping sustain conflicts is well known. However, the markets for these resources are often extra-local or global, such as for coltan or timber. Resources then present a theoretical problem for neoclassical economists, as they are meant to be a source of comparative advantage, but they are often associated with negative economic outcomes and conflict (Wengraft 2017). Thus, there is a kind of resource fetishism which takes places, which ascribes them both independent causative power and a derogation from the laws of economics. This obscures both the ways in which natural resources are socially produced, in part through their being part of global supply chains, and broader underlying social dynamics. Natural resources are socially produced because what gets defined as a resource shifts through time. For example, sand was not considered a resource for much of human history, but it is now a vital input into silicon chips, which make our electronic devices run.

Perhaps the most common explanation for underdevelopment, amongst Western publics at least, is the poor-country government corruption. There are two types of corruption: petty and grand (de Sardan 1999), both of which may be very damaging to development for a variety of reasons, including waste of resources and reductions in trust. Whereas World Bank theories argue that privatisation, for example, will reduce governing elites' possibilities for malfeasance and corruption, in many cases, privatisation deals have been used for corrupt ends to sell assets at below market prices to cronies, for example, who can then fund election campaigns (Tangri and Mwenda 2013; Wiegratz 2016).

Externalist theories, such as dependency or post-development, on the other hand, argue that the roots of underdevelopment lie in colonialism and subsequent discourses and practices of development. Assemblage thinking, through its focus on networks, disputes the separation of the internal and external. Rather, what is important is the developmental impact of often transnational assemblages or networks, although this is itself a heuristic, as ultimately everything on earth is connected to everything else, either directly or indirectly.

A new addition to externalist theories is the supposedly destructive nature of aid, which is seen by some, such as Dambisa Moyo, to distort the market. Whereas some argue that aid can correct for market failures, for extreme neoliberals such as Moyo or William Easterly, it is only through self-interest that societal welfare can be maximised, following Adam Smith's line of thinking (Wiegratz 2016). Thus, fraud and deception may become more commonly accepted under neoliberalism. Interestingly, however, William Easterly (2017) recently admitted on Twitter that Jeffrey Sachs was "more right" than him that free distribution of anti-malarial bed nets would dramatically bring down rates of infection in Africa.

As this book has suggested, development is a contested term without a stable meaning. Development can be used discursively to suppress dissent (Appadurai 2006) on the grounds that opposition groups are anti-development. Aid may have the (un)intended impact of supporting autocratic governments (Hagmann and Reyntjens 2016). There are also counter-tendencies. For example, it has been noted that as urbanisation brings attention to the political power distribution in a society, it can promote democracy (Dyson 2001).

According to Andrew Brooks (2017, 235),

> The 'end of development' does not mean that an end point for capitalist development is in sight, but rather that Western-led International Development and the vision of modernisation promoted by Europe and America is becoming a fading force in the Global South.

As China has assumed a leading role in the Global South, it has perhaps more actively promoted this concept, even as the relative roles of Europe and the US decrease. Thus, both immanent and imminent development look likely to continue into the future as both theories and practices. Designing effective policies and promoting international cooperation for sustainability are key challenges. The era of international development looks set to continue with new centres of leadership, such as China, and with new or renovated discourses around win-win cooperation. The fact that it is such a malleable and consequently successful concept, at least in terms of its own survival, diffusion and propogation, will be key to its continued longevity.

As Jeffrey Sachs (2018) has noted, "world capitalism will not get us on track". While the current system is very good at producing economic output, and this can be expanded at the cost of "human suffering" and "rapacious exploitation of the environment", there is a need for far greater equity and a shift towards sustainability. The idea of sustainable development then is a useful one, if it is conceived of as encompassing these ideals, through the SDGs, for example. However, equity and environmental protection are not intrinsic to the system. In fact, as Professor Sachs notes, the current globalised economic system has produced massive inequality and environmental despoliation. This suggests the need for stronger instruments of international regulation to achieve these goals and the development of instruments and policies to achieve these. The vested interests in the perpetuation of the current direction of the system are strong, and the fragmented nature of political authority across the planet makes coordination difficult, but not impossible. Important shifts are happening in renewable energy, for example. Building on these through education, activism, policy and transnational coordination consequently assume acute urgency. The nature of global development needs to be urgently reshaped if the promise inherent in the term is to be realised, rather than the actuality which has so far fallen short.

Note

1 Populism is defined by an appeal to "the masses" rather than particular classes, although it may be left wing or right wing. Examples of populists would include Donald Trump in the United States or Juan Perón, who was president of Argentina in the 1970s. Its rise in Western countries is associated with the so-called "elephant curve", which looks like an elephant and shows income growth per percentile of the global population. Incomes have been declining for the lowest percentiles and from approximately the sixtieth to the ninetieth percentile (which is composed in large part of the Western middle classes) for the last several decades.

References

Appadurai, A. 2006. *Fear of Small Numbers: An Essay on the Geography of Anger*. Durham, NC and London: Duke University Press.

Bond, P. 2006. *Looting Africa: The Economics of Exploitation*. Scottsville, South Africa: University of KwaZulu-Natal Press; London and New York: Zed Books; New York: Distributed in the USA by Palgrave Macmillan.

Boyce, J.K., and L. Ndikumana. 2001. "Is Africa a Net Creditor? New Estimates of Capital Flight from Severely Indebted Sub-Saharan African Countries, 1970–1996." *The Journal of Development Studies* 38 (2): 27–56.

Brooks, A. 2017. *The End of Development: A Global History of Poverty and Prosperity*. London: Zed Books.

Carmody, P. 2001. *Tearing the Social Fabric: Neoliberalism, Deindustrialization, and the Crisis of Governance in Zimbabwe*. Portsmouth, NH: Heinemann.

———., and P. Kragelund. 2016. "Who Is in Charge? State Power, Agency and Sino-African Relations", *Cornell International Law Journal*, 49, 1–24.

Collier, P. 2007. *The Bottom Billion*. Oxford and New York: Oxford University Press.

Dahi, O., and F. Demir. 2017. "South-South and North-South Economic Exchanges: Does it Matter Who Is Exchanging What and with Whom?" *Journal of Economic Surveys* doi:10.1111/joes.12225.

de Sardan, J. 1999. "A Moral Economy of Corruption in Africa." *Journal of Modern African Studies* 37 (1): 25–52.

Dell'Amore, C. 2014. "Has Half of World's Wildlife Been Lost in Past 40 Years?" *National Geographic*. http://news.nationalgeographic.com/news/2014/09/1409030-animals-wildlife-wwf-decline-science-world/.

Dyson, T. 2001. "A Partial Theory of World Development: The Neglected Role of the Demographic Transition in the Shaping of Modern Society." *International Journal of Population Geography* 7: 1–24.

Easterly, W. 2017. "Looks like @JeffDSachs Got it More Right Than I Did on Effectiveness of Mass Bed Net Distribution to Fight Malaria in Africa." https://twitter.com/bill_easterly/status/898606621361713153?lang=en.

Gibson-Graham, J.K. 1996. *The End of Capitalism as We Knew it: A Feminist Critique of Political Economy*. Minneapolis, MN: University of Minnesota Press.

Hagmann, T., and F. Reyntjens, (eds.) 2016. *Aid and Authoritarianism in Africa: Development Without Democracy*. London: Zed Books.

Internet World Stats. 2017. "Internet Usage Statistics: The Internet Big Picture: World Internet Users and 2017 Population Stats." www.internetworldstats.com/stats.htm.

Klein, N. 2014. *This Changes Everything: Capitalism Vs. The Climate*. London: Allen Lane.

Kragelund, P. 2014. "'Donors Go Home': Non-Traditional State Actors and the Creation of Development Space in Zambia." *Third World Quarterly* 35 (1): 145–62.

———., and P. Carmody. 2016. "BRICS' Impacts on Local Economic Development in the Global South: The Case of a Tourism Town and Mining Provinces in Zambia." *Area Development and Policy* 1 (2): 218–37.

Lall, S. 1992. "Technological Capabilities and Industrialisation." *World Development* 20 (2): 165–86.

Leyshon, A., and R. Lees, (eds.) 2003. *Alternative Economic Spaces*. London: Sage.

Navarro, P. 2007. *The Coming China Wars: Where They Will Be Fought and How They Will Be Won*. Upper Saddle River, NJ and London: Pearson FT Press.

———., and G. Autry. 2011. *Death by China: Confronting the Dragon – A Global Call to Action*. Upper Saddle River, NJ: Prentice Hall; London: Pearson Education [distributor].

Olusoga, D. 2017. "Empire 2.0 Is Dangerous Nostalgia for Something That Never Existed." *The Guardian*. www.theguardian.com/commentisfree/2017/mar/19/empire-20-is-dangerous-nostalgia-for-something-that-never-existed.

Oqubay, A. 2015. *Made in Africa: Industrial Policy in Ethiopia*. Oxford and New York: Oxford University Press.

Prahalad, C. 2009. *The Fortune at the Bottom of the Pyramid: Eliminating Poverty Through Profits*. Princeton, NJ: Pearson FT Press.

Rachman, G. 2016. *Easternisation: War and Peace in the Asian Century*. New York: Vintage Books.

Report of the High-Level Panel on Illicit Financial Flows from Africa. 2015. *Track it! Stop it! Get it? Illicit Financial Flows.*

Rosling, H. 2012. "Tweet." https://twitter.com/hansrosling/status/270834658630062080?lang=en.

Sachs, J. 2008. *Common Wealth: Economics for a Crowded Planet*. New York: Penguin Press.

———. 2018. *Opening Remarks, Sixth International Conference on Sustainable Development*. Columbia University, September 27th.

Sharples, N., T. Jones, and C. Martin. 2016. "Honest Accounts? The True Story of Africa's Billion Dollar Losses." www.francophonie.org/IMG/pdf/honest-accounts_final-version.pdf.

Stuenkel, O. 2016. *Post-Western World: How Emerging Powers Are Remaking Global Order*. Cambridge: Polity.

Tangri, R., and A. Mwenda. 2013. *A Politics of Elite Corruption in Africa: Uganda in Comparative African Perspective*. London and New York: Routledge.

Taylor, I. 2017. *Global Governance and Transnationalizing Capitalist Hegemony: The Myth of the "Emerging Powers"*. London and New York: Routledge.

Trump, D. 2013. "Every Penny of the $7 Billion Going to Africa as Per Obama Will Be Stolen – Corruption Is Rampant!" https://twitter.com/realdonaldtrump/status/351642052854951936?lang=en.

Van de Walle, N. 2001. *African Economies and the Politics of Permanent Crisis, 1979–1999*. Cambridge: Cambridge University Press.

Wengraft, L. 2017. "The Pillage Continues: Debunking the Resource Curse." *Review of African Political Economy*. blog. http://roape.net/2017/01/24/pillage-continues-debunking-resource-curse/.

———. 2018. *Extracting Profit: Imperialism, Neoliberalism and the Scramble for Africa*. Boston: Haymarket Books.

Whitwam. 2014. "$1200: The Price Of (Legally) 3D Printing Your Own Metal AR-15 Rifle at Home." *ExtremeTech*. www.extremetech.com/extreme/191388-1200-the-price-of-legally-3d-printing-your-own-metal-ar-15-rifle-at-home.

Whyte, W., and K. Whyte. 1988. *Making Montdragon: The Growth and Dynamics of the Worker Cooperative Complex*. Ithaca and London: Cornell University Press.

Wiegratz, J. 2016. *Neoliberal Moral Economy: Capitalism, Socio-Cultural Change and Fraud in Uganda*. London: Rowman and Littlefield.

Wood, R. 2017. *U.S. Imposing Anti-Dumping Duties on Chinese Aluminum Foil*. www.bloomberg.com/news/articles/2017-10-27/u-s-slaps-anti-dumping-duties-on-chinese-aluminum-foil-imports.

Index

Note: figures are denoted with *italicized* page numbers; note information is denoted with n and note number following the page number.

"This book builds some of the most important bridges between the intellectual work that so badly needs to be in the centre of our discourse on all aspects of development – global hunger, trade, development, urbanisation and the pre-crisis need to link economy, ecology and a global ethics. Its inter-disciplinary character will be particularly welcomed by students anxious to move from good scholarship into the most practical of applications, in both the short term and a hopefully sustainable future. Above all, I welcome it in its challenge to inevitabilities of thought and practice that need to be questioned, and it is a real contribution in that sense to the pluralist scholarship we need."

President Michael D. Higgins, Republic of Ireland

Taking a critical and historical view, this text explores the theory and changing practice of international development. It provides an overview of how the field has evolved and the concrete impacts of this on the ground on the lives of people in the Global South.

Development Theory and Practice in a Changing World covers the major theories of development, such as modernisation and dependency, in addition to anti-development theories such as post-modernism and decoloniality. It examines the changing nature of immanent (structural) conditions of development in addition to the main attempts to steer them (imminent development). The book suggests that the era of development as a hegemonic idea and practice may be coming to an end, at the same time as it appears to have achieved its apogee in the Sustainable Development Goals as a result of the rise of ultra-nationalism around the world, the increasing importance of securitisation and the existential threat posed by climate change. Whether development can or should survive as a concept is interrogated in the book.

This book offers a fresh and updated take on the past 60 years of development and is essential reading for advanced undergraduate students in areas of development, geography, international studies, political science, economics and sociology.

Pádraig Carmody is Associate Professor in Geography, Head of Department, Fellow and director of the Masters in Development Practice at Trinity College Dublin and Senior Research Associate at the University of Johannesburg. His research centres on the political economy of globalisation and economic restructuring in Southern and Eastern Africa. He has published in a variety of journals, such as *Economic Geography*, *World Development* and *Political Geography*, amongst others. He has published seven books. The second edition of his *New Scramble for Africa* has recently been published. He sits of the boards of the *Journal of the Tanzanian Geographical Society*, *African Geographical Review*, *Political Geography* and *Geoforum*, where he was previously editor-in-chief. He is currently an associate editor of the journal *Transnational Corporations*, published by the United Nations Conference on Trade and Development.

DEVELOPMENT STUDIES / DEVELOPMENT THEORY

Cover image: © Getty Images

www.routledge.com

Routledge titles are available as eBook editions in a range of digital formats

an **informa** business

ISBN 978-1-138-55178-7

9 781138 551787